Enchantment by Birds

Russell McGregor has been a birdwatcher since childhood and an historian for over 40 years. Formerly a professor of history at James Cook University, he has written numerous articles on birds and birdwatching, and is also the author of six previous books, including *Indifferent Inclusion: Aboriginal people and the Australian nation* (winner of the 2012 New South Wales Premier's Australian History Prize), and *Idling in Green Places: a life of Alec Chisholm* (shortlisted for the 2020 National Biography Award).

Russell McGregor

Enchantment *by* Birds

a history of birdwatching in 22 species

SCRIBE

Melbourne | London | Minneapolis

Scribe Publications
18–20 Edward St, Brunswick, Victoria 3056, Australia
2 John St, Clerkenwell, London, WC1N 2ES, United Kingdom
3754 Pleasant Ave, Suite 100, Minneapolis, Minnesota 55409, USA

Published by Scribe 2024

Typeset in Portrait by the publishers

Printed and bound in Australia by Griffin Press

Scribe is committed to the sustainable use of natural resources and the use of paper products made responsibly from those resources.

Scribe acknowledges Australia's First Nations peoples as the traditional owners and custodians of this country, and we pay our respects to their elders, past and present.

978 1 761381 44 7 (Australian edition)
978 1 957363 97 4 (US edition)
978 1 761386 01 5 (ebook)

Catalogue records for this book are available from the National Library of Australia and the British Library

scribepublications.com.au
scribepublications.co.uk
scribepublications.com

To Christine, Caitilin, and Lachie, with love.

Contents

Preamble: Capricorn Yellow Chat

A BRILLIANT LITTLE YELLOW gem flashed onto a twig by the roadside up ahead. It preened briefly and then flew across the road, where it flitted among the samphire, its luminescent saffron feathers glowing against the grey-green foliage. It was a Capricorn Yellow Chat, a bird for which we'd been searching for two days. We'd trudged up and down the muddy roadside, past the saltworks with their blinding-white crystal mountains, dodging the huge trucks that thundered down the road from nearby Port Alma at the mouth of the Fitzroy River. We'd squelched into swamps where sandflies vied with mosquitoes for which one could deliver most itch per square inch of skin. We'd seen lots of birds, but the chat had eluded us until we went back to the spot, near Cheetham's Saltworks on the Port Alma Road, where we had begun our quest two days before. And there it was, soon joined by another, both flaunting their beauty against the starkness of the saltpans.

Like other birders, I'm exhilarated by seeing a species I haven't seen before. Like other birders, too, I enjoy the challenge of tracking down a bird, especially rare ones, as Capricorn Yellow Chats are. The thrill of the chase is among the pleasures of the pastime. So is aesthetic appreciation, and the chat is a visually stunning bird. Famed field guide author Peter Slater marvelled at 'the almost painful luminosity of the male's colouring, a yellow so glowing and pure that the artist must despair of reproducing it'.[1] Admittedly, in the other avian aesthetic domain—song—it is distinctly mediocre. Scientifically, the Yellow Chat is fascinating for several reasons, including its disjunct distribution, with the Capricorn subspecies inhabiting isolated patches of coastal saltmarsh in central Queensland, hundreds of kilometres from its nearest kin in the dry west of that state.

The Capricorn Yellow Chat has an intriguingly tangled history, too.[2] Its first recorded encounter with the British colonisers of Australia was in 1859, when publican and taxidermist John McGregor shot a little yellow bird somewhere near Rockhampton. He didn't know what it was, so he skinned it and sent the specimen to the National Museum in Melbourne. However, science's near acquaintance with the chat at this point proved abortive. For reasons that remain unclear, the museum staff failed to appreciate the significance of the skin they'd been sent, and did nothing with it, even though it was a bird as yet undescribed by science. McGregor's bird skin lay neglected in the museum for the next 100 years, its scientific significance unrecognised.

Meanwhile, the Yellow Chat as a species came to the attention of scientists via a different route and a different subspecies. In 1877, Edward Pierson Ramsay, curator of the Australian Museum in Sydney, obtained a skin from a naturalist named Thomas Gulliver, who had shot the bird near Normanton in Queensland's Gulf Country. Ramsay did appreciate the significance of what he had been sent. Together with the Melbourne-based Consul General de France, le compte de Castelnau, he presented a full description of the Yellow Chat, *Epthianura crocea*, to an 1877 meeting of the Linnean Society of New South Wales. Thus the Yellow Chat entered the realm of science bearing a specific name paying tribute to its most striking feature: its bright saffron feathers.

Thereafter, Yellow Chats were occasionally seen near Rockhampton—only occasionally, because they're so rare and so scattered. The visiting Norwegian zoologist Carl Lumholtz shot three at Fitzroy Vale near the mouth of the Fitzroy River in 1882, and sent the skins back to a university in Norway. In 1917, the prominent amateur ornithologist A.J. Campbell mentioned that two of his birding colleagues had recently taken a Yellow Chat specimen at Torilla Plains, north of Rockhampton.[3] It was just a passing mention, after which there were no more records of Yellow Chats in central Queensland for 75 years.

In 1958, Australian ornithologist Allen Keast identified the central Queensland Yellow Chat as a distinct subspecies. He named it *Epthianura crocea macgregori*, commemorating the man who in 1859 had collected the first specimen, which had been gathering dust in the museum ever since. However, he misunderstood where John McGregor had collected his specimen, thinking it was well inland of Rockhampton, so he gave the subspecies the misleading vernacular name Yellow Chat (Dawson). Perhaps the misunderstanding is understandable, since Keast was describing a subspecies that had not been seen for 50 years. The bird to which he assigned a taxonomic status might well have been extinct. Two decades later, Peter Slater suggested it was.[4]

But the Capricorn Yellow Chat was one of many Australian bird species and subspecies that have pulled back from the brink of extinction—or at least pulled back far enough to not yet fall over the edge. In 1992, Don Arnold, Ian Bell, and Gary Porter rediscovered the Capricorn Yellow Chat on Curtis Island at the mouth of the Fitzroy River. Bell recalled the moment of rediscovery with laconic humour:

> "what's that bird Don?" "I dunno" and after looking in Simpson and Day, (although I think Don was a Slater's man) Don announced ... "it's a Yellow Chat" to which I said ... "hmm should that be here?" to which Don said "nup, occurs around western Queensland bore drains" to which I said "well bugger me!" or something along those lines.[5]

Boosted by the rediscovery, ornithologists searched other likely locales, finding the chat at Torilla Plains in 2003, Twelve Mile Creek near Marmor in 2004, Fitzroy Vale in 2008, and several other central Queensland sites over subsequent years. Numbers were tiny, and the subspecies was declared Critically Endangered. But it had survived, and continues to do so, drawing a constant stream of scientists to study it, birdwatchers to admire it, and twitchers to tick it off.

A feature of the foregoing narrative that may strike readers today is the amount of shooting involved, especially during the chat's early encounters with European people. In the 19th century, and well into the 20th, a gun was a normal part of a birder's kit, and shooting birds, both to identify them and to souvenir their skins, was commonplace. Before the development of high-precision optical equipment, having a bird in the hand was often the only way to clinch an identification; and collections of skins were the foundations on which ornithological science was built. Shooting and skinning birds was not the preserve of professional ornithologists (who were then so few in Australia that they could be numbered on the fingers of one hand), but an everyday practice of the amateurs who dominated bird study. One of the themes I'll tease out in this book is the displacement of shooting by less bloody means of bird appreciation.

The decline of shooting was related to, and partly driven by, improvements in binoculars and cameras, plus the advent and increasing sophistication of field guides. Those developments will be explored in this book, as will birding's perennial connections with conservation. Even in the heyday of shotgun ornithology, the protection and preservation of birdlife were prominent goals of the birding fraternity; and with later changes in science and the wider culture, the conservationist component of birding blossomed into an ecological imperative to safeguard the diversity of our avifauna. Conservation, like birding itself, has emotional as well as rational and pragmatic strands, and the intertwining of those is among the themes I examine. I can't offer a comprehensive history of Australian birdwatching in one short book, but I hope to illuminate what has impelled birding in the past and still impels it today. It's a story in which continuities are as evident as changes.

A topic to which I devote some attention is bird names, particularly vernacular names. Naming, after all, is a kind of bedrock of birding. The first thing that novice birdwatchers have to learn is which name to pin to the fluttery, feathery thing they can barely hold in binocular focus. Getting the name right may be even more important for the seasoned old hand, whether for pride or to prove proficiency. Beyond that, many birders have an emotional investment in bird names and an attachment to those names they consider apt or appealing. Name changes have raised the hackles of birders in the past—for example, when, in the late 1970s, 'Stone-curlews' became 'Thick-knees', and the 'Jabiru', a prosaic 'Black-necked Stork'. Other name changes have quickly gained the tick of approval, as when the clumsy 'Yellow Chat (Dawson)' gave way in 2004 to the more euphonious 'Capricorn Yellow Chat'. As those examples suggest, the aesthetics of naming is a major part of the process.

Indeed, aesthetics pervades birding in all its aspects, most of all at the primary point of perception. That's why Neville Cayley began Australia's most famous field guide, *What Bird Is That?*, published in 1931, with a paean to the beauty of birds and the emotions they arouse in us:

> Birds express all that is beautiful, joyous, and free in nature. They delight our eyes, charm our ears, quicken our imagination, and through association with the bushland inspire us with a profound love of country.
>
> What visions of freedom and joy come to us when we see a flock of Scarlet Honey-eaters feeding among the blossoms of a tea-tree; a Spinebill sipping nectar from a native fuchsia; a Blue Wren moving among the golden beauty of a wattle-tree; or Silver Gulls flying lazily above the limpid blue waters of our harbours? What pleasure is ours when we hear the joyous carefree carolling of Magpies at dawn; the springtime song of the Grey Thrush; the wonderful song mimicry of the Lyretail, or a Song-lark soaring heavenwards filling the air with its melody.[6]

A field guide author of a later generation, Peter Slater, put it more succinctly: 'Birds to me are feathered poems.'[7]

As Cayley's and Slater's words attest, the aesthetics of birding encompass far more than admiring a pretty bird—although bright colours and a perky personality can make admiration easier, as the Capricorn Yellow Chat testifies. Yet a bird does not have to be conventionally pretty to be aesthetically appreciated. It may be how the bird flies: White-throated Needletails scything the air with sickle wings, or Wedge-tailed Eagles soaring skyward with the barest tilt of tail and wing, are stunning sights. It may be how the bird feeds: the aerial acrobatics of a Grey Fantail, or the ground-focussed fussing of a Brown Quail, or the deep-diving dexterity of a Hardhead are actions of rather plainly plumaged birds, but entrancing nonetheless. We might focus on any of numerous facets of birdlife, from calls to courtship, from nesting to anting: all have their charm and allure. Much of the aesthetic attraction of birdwatching comes from appreciating not birds alone, but rather birds as denizens of nature.

Fundamentally, birdwatching is a means for people to connect with nature. More than that, it's a manifestation of modern, urban people's craving for communion with nature, driven in part by an uneasy feeling that in this mechanised and technologised world we've lost connection. That's not birdwatching's only motivation. Beyond it lie a whole muddle of motives, from a thirst for knowledge to a love of lists, from the twitcher's competitive compulsion to score the biggest tally of ticks to the aesthete's exultation in birdsong at dawn. Yet connecting with nature is, and always has been, a powerful impetus behind birding. Field guide author Graham Pizzey knew this, describing birdwatching as 'an escape to reality, an escape which puts you in touch with the stability, integrity, and reassurance of the natural world'.[8] A glossy advertisement on the back cover of *Australian Birdlife* magazine expresses much the same sentiment, proclaiming Swarovski binoculars as instruments that allow birders to become 'One With Nature'. Similarly, the publisher's blurb for Paul Sorrell's *Getting Closer: rediscovering nature through bird photography* claims that the book 'offers a simple, practical path for readers to begin to "rewild" themselves'.[9]

Birding puts us in touch with the wild. By the 'wild', I do not mean what some call 'wilderness': a politically fraught term that locates a mythically pristine nature somewhere beyond the taint of humanity. The wild, by contrast, is all around us. It's the world of living things that humans don't directly control: the plants and animals that are not domesticated but that live unrestrained, though not unaffected, by people. The wild is a domain of wonder, outside humanity and yet our tangibly close companion. By giving access to the wild, birding allows us to reach beyond ourselves and beyond the confines of human society, into a world we never fully comprehend and yet can appreciate both emotionally and cerebrally. Birding is not an anti-social activity (although some enthusiasts prefer to practise it so), but one that alerts us to the interfaces between humans and animals—or between culture and nature—thereby enhancing our understanding of ourselves as inhabitants of this planet.

Seeing the Capricorn Yellow Chat was an encounter with the wild. It certainly was not a wilderness experience. We saw the chat from the edge of a sealed road that vibrated underfoot as motorised behemoths pounded by. The backdrop juxtaposed mountains of salt against neatly rectangular evaporation pans and not-so-neat piles of rusting machinery. But the chats—and other birds, for Mangrove Honeyeaters, Australian Pipits, and Red-capped Dotterels were there in abundance, too—made it an experience of the wild. That's what birds bequeath to us.

One of the wonderful things about 'rewilding' by birding is that birds are so accessible. They can be found everywhere on the planet except the extreme South Pole, and even there they sometimes cruise by. That 'everywhere' includes urban and near-urban places. To find birds, you don't need to journey to remote locales; you can find them close by home. As one of Australia's greatest birdwatchers, Alec Chisholm, explained to the readers of Sydney's *Telegraph* in 1926, they did not have to 'crawl into all manner of queer corners in order to see birds at their best'. The joys of birding were easily accessible, he enthused, and could rejuvenate the souls of city people. In his characteristically lavish style, Chisholm lauded birdwatching as a means to 'open the spiritual eye that develops when you come to regard the flash of a wing or the snuggling of a small mother on a nest as one of the most gracious things outside Utopia'.[10]

I've attributed birdwatching to modern, urban people's yearning for connection with nature. Obviously, however, people watched birds long before they became modern and began living in cities. They've always watched birds. Australia's Indigenous peoples have watched the birds around them from time immemorial, doubtless for pleasure as well as myriad other motives, practical, spiritual, and cultural. But while watching birds has been ubiquitous in human history, birdwatching as a pastime is of much more recent vintage. As American and British historians of the hobby have shown, birdwatching emerged in the English-speaking world toward the end of the 19th century as a form of edifying recreation that evolved from the popular natural history of the Victorian era.[11] Birdwatching then was not the same as birdwatching now, but there are sufficient continuities between its practice then and now, and sufficient divergences from what went before, to date birdwatching's beginnings to the last decades of the 1800s.

Although this is a history, I haven't structured the story chronologically, but rather by birds, each chapter taking its title from the name of an avian species and a theme connected with that species. (Actually, this preamble takes its title from a subspecies, but I trust that small anomaly can be forgiven.) Some chapters cleave close to the bird of the title; many meander further afield; most emulate the opportunistic practices of birders themselves, chasing new species and themes whenever they come into view. Each chapter includes some discussion of the bird nominated in the title: its habits, appearance, calls, quirks, and so forth. But most of my words are not about the corporeal qualities of actual avifauna, instead exploring historical connections between birds and people. All birds bear long plumes of historical association, residues of the myriad ways in which the lives of birds and people have intertwined and intersected.

My own life and that of the Capricorn Yellow Chat intersected in a way I learned about shortly before I saw the bird on the Port Alma Road. John McGregor, who shot the first, long-neglected specimen, was my great-great-grandfather. I don't know much about him. According to family lore, he left his home in Northern Ireland in the 1850s, his forebears having been evicted from the Scottish Highlands in the clearances a few generations earlier. Like most who came to Australia in that decade, he sought his fortune in gold. He found it in grog, running the Ulster Arms and Union hotels in Rockhampton. He wasn't a birder, at least not in today's understanding of that term. But he was an entrepreneur and a skilled taxidermist, in those respects resembling the most famous birdman of the era, the Englishman John Gould. My great-great-grandfather had none of the eminence of Gould nor his expertise, but he did have an interest in the birdlife around him, so when he saw a bird that seemed new to him, he tried to identify it by the only means then available.

John McGregor took a Capricorn Yellow Chat with a shotgun in 1859. One hundred and sixty years later, I looked at one through a pair of 8x42 Swarovski binoculars while my birding mate, Vince, fired off 20 frames per second with a Sony A9 camera fitted with a 200-600mm zoom. Between us, we probably had as detailed a view of the bird as my great-great-grandfather had when he held that first specimen in his hand. That's testament to the technological changes that have helped transmute birding into a bloodless form of nature appreciation. Yet myriad other factors—emotional, attitudinal, cultural, scientific, and institutional—have impelled birding in the same direction while simultaneously pushing it in others. The following chapters explore those myriad directions.

Terminological tangles

In the preamble above, I've tried to convey some of the motivations and exhilarations of birding and to signal what comes in the pages ahead. Along the way, I've used some words that are overdue for clarification: 'birdwatching', 'birding', 'ornithology', and 'twitching'. These are slippery words, and while they can never be precisely pinned down, they need to be clarified to the point that my intended meaning is apparent.

'Ornithology' could be taken to mean the scientific study of birds; 'birdwatching', the recreational appreciation of birdlife; and 'twitching', the assiduous ticking-off of birds new to the observer. The trouble is, the categories continually collapse into each other. Recreational birdwatchers commonly contribute to scientific ornithology, and have done so since birdwatching's beginnings. Ornithological scientists often spend their leisure hours enjoying a spot of recreational birdwatching. Twitchers, most of the time, turn out to be much like ordinary birdwatchers, and sometimes they're professional ornithologists as well. There are no clear demarcations or rigorous divides between these categories—indeed, I doubt whether they can properly be called 'categories' at all.

The words 'birding' and 'birder' offer a convenient way around the tangle, and I use them as catch-all terms to designate all forms of interest in wild birds and all who pursue those interests. The other terms are still useful, because it's often important to know where the emphasis was placed in a particular instance of birding. So in the pages ahead, I use 'ornithology' for enterprises with an emphasis on science, 'birdwatching' where recreation was a prime consideration, and 'twitching' where the stress was on ticking off new sightings. But the differences are of emphasis only, and the words should not be taken to refer to tightly bounded categories of bird appreciation. 'Birding' encompasses them all.

CHAPTER ONE

Galah: Beauty

THE IDEA FOR THIS book first came to me at the National Library in Canberra. Not only are the ornithological riches of that institution almost overwhelming; so is the abundance and diversity of birdlife in that city. Canberra must be the most bird-blessed national capital in the world, and among its most plentiful and beautiful birds are parrots and cockatoos. On 16th-century maps, Australia was called *Terra Psittacorum*, or the Land of Parrots. Following that line, Canberra could be called the capital city of parrots.

Crimson Rosellas abound not only in Canberra's leafy suburbs, but also in the bustling CBD. Eastern Rosellas and King Parrots, while less abundant, are still common. Red-rumped Parrots flash by almost everywhere, while Superb Parrots, aptly named for their resplendent green plumage topped with a scarlet throat and yellow face, are increasing in number. So are Rainbow Lorikeets, adding their shrill shrieks to the harsher screeches of the more numerous Sulphur-crested Cockatoos and Little Corellas. Gang-Gang Cockatoos, recently declared endangered in New South Wales, are common garden birds in Canberra.

Then there are the Galahs. They're exquisite birds: not gaudy like some parrots; nor almost monochrome like many cockatoos. The Galah's beauty is subtle, resting on a delicate combination of rose pink and lavender grey, capped off with a short, pale crest. In Canberra, hundreds of these beautiful birds can be seen feeding in parks and gardens, or wheeling colourfully overhead while giving their distinctive high-pitched contact call.

It was not always so. When building the national capital began in the 1910s and 1920s, no Galahs lived there. By the time they arrived, a couple of decades later, the former sheep paddocks had been transformed into buildings and boulevards, suburbs and streets. The Galahs loved it. Joseph Forshaw, an ornithologist specialising in parrots, observed in 1969 that, 'In little more than two decades the Galah has built up its numbers so rapidly in the Canberra district, where it was formerly a rare visitor, that it has now become one of our most common birds.'[1]

It was a similar story across the continent. When Australia was first colonised, Galahs were confined to the interior and north, and there they were not particularly plentiful. Historian Bill Gammage has documented the spread of the species, which began in the 19th century and accelerated in the 20th, so today they're common all over Australia except in the wettest rainforests and the driest deserts. Gammage attributes the expansion of Galahs, in both range and numbers, to the disruption of Aboriginal land-management strategies, particularly the demise of an Indigenous economy in which yams and tubers were staple foods. It was sheep- and cattle-grazing in the first instance, he argues, that created the environmental conditions that launched Galahs on their successful invasion of the continent.[2] Whether or not one accepts the finer detail of Gammage's carefully crafted argument, his fundamental point that Galahs increased in consequence of a shift from Aboriginal to European modes of environmental management is surely true.

Galahs are not the only avian species to have benefitted from that shift. Other familiar beneficiaries include Crested Pigeons and Little Corellas, once confined to the dry inland, but now abundant in coastal cities, parks, and gardens. Unfortunately, more birds have suffered from the shift, their ranges contracting and numbers diminishing. Emu and Brolga are two examples, still well-known, but now seldom seen by most Australians. Of course, attributing the over-arching process of species' expansion and contraction to a shift from Indigenous to European land-management practices does not preclude other factors. Doubtless, some wild Galahs are descendants of aviary escapees; some were deliberately released; some followed railway lines bordered with spilled grain until they reached the bonanza of the wheatfields. But for a species to thrive where once it did not requires environmental transformation of some kind; and in recently colonised lands like Australia, such transformations can usually be traced to the colonisers.

Despite its beauty, the spread of the Galah was often unappreciated by those responsible for its expansion. Galahs were considered pests in wheat-growing regions, and as they spread into those areas, they were shot and poisoned *en masse*. A prominent Australian conservationist of the mid-20th century, Crosbie Morrison—the 'Voice of Nature', as he was dubbed for his popular radio shows—prefaced an article on the spread of the species into South Australia's wheatfields in the 1940s with a note explaining that 'the depredations of flocks of galahs have forfeited for them the protection which they might have been accorded on account of the beauty of their coloring of grey and sunset pink'.[3] Decades later, ornithologist Stephen Debus similarly suggested that the

Galah's reputation as an agricultural pest has inhibited the popular appreciation of its beauty.[4]

That may have been so, especially in Morrison's day, but I doubt whether many Australians today care what Galahs might do to farmers' crops. Presumably, the farmers themselves care, but most urban Australians seem indifferent to such things. Yet the beauty of Galahs is still often underappreciated.

That's probably because Galahs are so common. Among the many who have made that suggestion was the eminent ornithologist Dom Serventy, who considered Galahs 'perhaps the loveliest' of all parrots and cockatoos.[5] Superlatives aside, there's some truth to his claim that familiarity breeds underappreciation: that commonness diminishes our capacity to recognise beauty. It's true not only of Galahs. Rainbow Lorikeets provide an equally striking instance of the splendour of a common bird eliciting mere passing admiration.

That's not to say that the beauty of common birds is ignored. It is *under*appreciated, not *un*appreciated. Yet while beautiful common birds may be admired, rarity somehow lends a special gloss to a bird's gorgeousness. Birders are prone to such preferential perceptions even though, at another level, they may aesthetically appreciate all birds. Aesthetics are crucial to birding, but, in the mix of motives and aspirations that impel the pastime, the admiration of beauty can be compromised by a craving for novelty. Birders admire Galahs and have done so throughout the pastime's past; yet the much rarer Major Mitchell's Cockatoo, garbed in more subdued shades of pink and white, will much more likely get a birder's heart racing.

Overseas visitors, who usually care little about a bird's local commonness, typically see things differently. When the great American field guide author Roger Tory Peterson first visited Australia in December 1965, he specially asked his host, Graham Pizzey, to show him a Galah. Pizzey obliged, showing the American a whole flock of Galahs coming to drink at an earthen dam at sunset on the Riverina plains. Peterson was entranced by their beauty, and eagerly recorded the scene. 'I don't know how much film raced through Dr Peterson's camera', Pizzey remarked, 'but I'm sure it will delight his American lecture audiences.'[6]

For many birders, it was the beauty of birds that first lured them into the hobby. The British evolutionary biologist Julian Huxley fondly recalled his own induction into birdwatching around the age of 14, prompted by the sight of a Green Woodpecker at his aunt's country house:

> I saw every striking detail: the rich green of the wings, the flash of bright yellow on the back when he flew, the pale glittering eye, the scarlet nape, the strange moustache of black and red; and the effect was as if I had seen a bird of paradise, even a phoenix. I was thrilled with the sudden realization that here, under my nose, in the familiar woods and fields, lived strange and beautiful creatures of whose strangeness and beauty I had been lamentably unaware.

'Most bird-watchers', Huxley continued, 'are started off on their hobby by some such sudden glory.' He went on to explain that the 'next step in their career is generally the same new realization of beauty and strangeness in other common birds; and only then, as familiarity dulls the edge of novelty, do they turn to look for rarities to give them new excitement'.[7] Huxley wrote those words in 1930, but they are as true today as they were then—indeed, truer in an age of technology, travel, and twitching.

Questing for something rare or unusual gives zest to birding, but an underlay of unassuming appreciation of beauty endures. Commonness may dull appreciation to some extent; rarity may sharpen admiration, but finding beauty in birds still motivates people to don binoculars and enter the avian domain. The beauty they seek is more than feather deep. It extends to birds' behaviour, their demeanour, and their mannerisms.

Like many species, Galah couples engage in mutual preening (allopreening), where each bird grooms those patches of its partner's plumage, such as face and neck, that the individual finds difficult to reach. This typically entails a good deal of face-nibbling that from a human perspective looks like kissing, or even sexual overtures. Reinforcing that impression, allopreening is part of Galahs' courtship repertoire, along with caressing and locking beaks: behaviours that readily prompt human analogies. Scientists tell us that, for Galahs as for other species that allopreen, the practice serves to reinforce the pair-bond. From a layperson's perspective, it looks like an expression of affection. Perhaps that comes to much the same thing. In any event, people find such human-like behaviours endearingly attractive.

They find birds' ability to transcend human limitations even more attractive. At ease in the heavens, fluttering and floating on thin air, birds in flight hold a magical appeal to earth-bound humans. That's not uniquely a birdwatcher's perspective. People have always appreciated the magic and magnificence of birds in flight, granting them a special place in our hearts and imaginations, in our art and literature, even in our religions. Birdwatching as a hobby merely offers a modern way to come under the avian spell.

Individual birds flying—a kestrel hovering above a field, for example, or a pelican planing down to water, or a songlark scaling vertically skyward—have a beauty of their own, but flocks in flight can be especially impressive. Galah flocks, alternating their colours in synchronised flight, are breathtakingly beautiful, as John Leach expounded in Australia's first field guide, *An Australian Bird Book*, in 1911:

What is more glorious than a mob of Rose-breasted Cockatoos (Galahs), 500 strong, airing their beauties and graces as they take a constitutional before retiring for the night? Probably no other kind of bird shows better company-flying than Galahs; now one sheet of delicate gray lavender, and the next instant a flash of brilliant salmon-pink, as the whole company turns and wheels, obedient to some command or signal unperceived by us; again, the sun lights up the pale-pink crests and gray backs, as they turn once more and wheel, screeching, to continue their evolutions further afield.[8]

Leach's appreciation rested on the changing hues as well as the gracefulness of 'company-flying', but even without the colour, coordinated flight is awe-inspiring. A murmuration of starlings, with thousands flowing across the sky in ever-changing geometric configurations, provides an apt example. There's no visible colour but black, yet the shape-changing spectacle is stunning.

The beauty of flight—and avian beauty more generally—can be fully appreciated only in birds that are wild and free. We need not go to the extremes of the 18th-century visionary poet William Blake—according to whom, 'A robin redbreast in a cage/ Puts all heaven in a rage'—to see in wild birds a gracefulness and beauty that birds behind wire lack. Charles Barrett, writing in 1907, understood this. The Diamond Firetail is 'one of the most beautiful of Australian birds', he enthused, 'and it is delightful to watch their graceful fluttering' as these 'dainty little creatures flit about in small flocks above the seeding grasses'. 'They are favourite cage birds', he added, 'but lose half their beauty when behind the prison wires. No wild creature is seen to advantage in captivity; you must make its acquaintance in its native haunts.' Like other birders, Barrett had a holistic appreciation of birds in nature, explaining that 'the bird, with the rolling plain and the blue sky, is part of an harmonious whole, and if you remove it from the wild half the charm is gone'.[9]

Half the charm, he specified, not all of it. Birds behind wire can still be beautiful, and birders can still appreciate their charms. Some birders have been aviculturists, most notably, perhaps, Neville Cayley, who as well as authoring Australia's most famous field guide also published *Australian Finches in Bush and Aviary* (1932) and *Australian Parrots: their habits in the field and aviary* (1938). Both the latter works give extensive advice on keeping finches, parrots, and cockatoos in cages; they express solicitude for the welfare of those 'lovely captives'; and they extoll aviarists' contributions to conservation. Yet Cayley's admiration of avian beauty in captivity was counterbalanced with some misgivings.

As he was acutely aware, beauty could be to a bird's detriment. For their exquisite attractiveness, birds such as Gouldian and Painted Finches, and Scarlet-chested and Turquoise Parrots, were trapped in vast numbers to feed the aviary trade. By the 1930s, the trapping and trafficking of birds were restricted by law, although Cayley and his fellow birders feared that lax enforcement was leading toward 'the extermination of some of the loveliest of our birds'. 'If the present wholesale trafficking continues', he wrote of the pastel-plumaged Bourke's Parrot, 'it will not be long before this lovely parrakeet will be completely wiped out of existence.'[10] Extinction has always haunted birders' imaginations, and the idea that the process could be propelled by a bird's beauty made it doubly abhorrent.

Birds were not only captured, but also killed for their beauty. Reporting on the Forest Kingfishers that visited his camp in north Queensland in 1886, A.J. Campbell wrote that he 'felt a reluctance to shoot them, but the exquisite beauty of their lovely jackets of Prussian blue and snow white vests were too tempting, and we bagged a few brace to preserve their skins for posterity'.[11] Campbell collected skins for ornithological study, but beautiful birds met a similar fate for domestic decoration. Stuffed and mounted, in glass domes and cases, dead birds were common ornaments in 19th- and early-20th-century homes and pubs. Colourfully attractive varieties such as kingfishers and parrots were among the most highly prized, but others suffered for their beauty, too. In 1910, Frank Littler attributed the rarity of the White Goshawk to 'the fact that whenever it appears close to civilization violent efforts are made to shoot it, on account of its handsome appearance'.[12] A White Goshawk must have made a stunning adornment to the mantlepiece.

A.J. Campbell in 1900 described the White Goshawk as 'a most beautiful creature' that embodied the awesome aesthetics of birds of prey. In a Tennysonian revelling in 'nature red in tooth and claw', he imagined how 'Some beautiful pictures may be imagined of this Goshawk in snow-white dress, with its dying quarry held in its relentless grip—perhaps a Parrot of many gay colours, or perchance a male Regent Bird in its handsome black and golden garb.'[13] Later birders expressed it differently, but the aesthetic adoration of raptors remains strong. Peter Slater extolled the stoop of the Peregrine Falcon as 'the most inspiring and beautiful sight to be experienced in the natural world'.[14] David Hollands described a Brown Goshawk's attack on a murmuration of starlings as 'a drama of great beauty'.[15]

A raptor's razor-edged beauty prompts one kind of aesthetic appreciation of birds. At the other end of the appreciation spectrum is avian clowning, and Galahs are among the most accomplished of clowns. Encountering hundreds of 'exceedingly clamorous Galahs' near Mataranka in the Northern Territory in the early 1960s, birder and journalist Michael Sharland enjoyed their 'fascinating acrobatics on the telephone wires':

I once thought that their contortions were due to the wires being slack and the consequential difficulty of holding on, but on several occasions later I found they would perform, evidently for the joy of it, and just as readily, on the fine, rigid branches of a dead tree. Here I watched them against the sky, spreading their wings and yelling, waving their heads about with stumpy crests erect, and sometimes turning a complete circle upside down with their feet clinging to the branches. It was quite an art this turning turtle and coming back to the upright again, and they did it purely in fun I believe.[16]

Watching birds having fun brings joy to people. Beyond that, it reinforces a sense of kinship with them.

Birders have long celebrated that sense in their descriptions of birds at play. In 1911, for example, A.H.S. Lucas and W.H.D. Le Souëf delighted in how Magpies expressed their 'excess of vitality ... in sundry gambols and mirthful frolics, such as running round trees or stumps as if playing at hide and seek; darting at one another as if pretending to want to fight, or trying to catch one another by the tail when on the wing'.[17] At times, some scientists have looked askance at the notion that birds do things just for fun, apparently from an exaggerated fear of anything that smacked of anthropomorphism. Most birdwatchers, however, have happily accepted that birds play for the sheer joy of playing, and some have pushed hard the idea that having fun is a significant motivator of bird behaviour.[18] Recent scientific research lends support to that contention.[19]

The same research indicates that birds possess a high level of intelligence. Corvids are generally ranked the most intelligent birds, but cockatoos come high on the scale, too. Again, this has long been recognised, Lucas and Le Souëf remarking in 1911 that there 'can be no question of the very high degree of intelligence' possessed by cockatoos.[20] One hundred years later, ethologist Gisela Kaplan backed up that assessment in a survey of scientific studies demonstrating birds' proficiency at complex cognitive tasks. Cockatoos were among the most proficient. Kaplan's 2015 book, *Bird Minds*, was inspired by her 75-year-old pet Galah still showing a keenness to learn.[21] Recognition of a bird's intelligence prompts feelings of mutuality, and Galahs are adept at that.

Galahs, Dom Serventy remarked in 1969, are among those birds 'which seem to take friendly personal interest in the watcher. Their drollery as they peer, almost in a human-like affectionate way, at an observer, adds to their appeal'.[22] Seeing a Galah eyeing us, inquisitively rather than fearfully, bonds us with the bird and boosts our appreciation of its beauty. But what does the Galah see? I doubt it sees us as beautiful. Humans are much too common for that. We'll never know quite what the bird makes of us. Galahs can be taught to talk, but that endears them to us rather than enabling communication between us. Galahs—like all birds—will remain forever other to we humans; that's part of their attractiveness. Yet we can gaze across the species gap in a spirit of fascination and exploration—reciprocally in the case of the Galah. I'm sure they don't people-watch in the same way as we birdwatch, but the fact that at least some species peer back at us with interest adds immeasurably to the allure of birds.

CHAPTER TWO

White-throated Gerygone: Melody

IN AUSTRALIA'S FIRST BIRDING field guide, published in 1911, John Leach called the White-throated Flyeater a 'Musician'.[1] That single word was his only intimation that this bird is a gloriously gifted songster. Admittedly, bird calls were not strong suits in Leach's *Australian Bird Book*. For many species, he gave no indication at all of vocalisations; and for those he did, the descriptions of calls and songs were usually too terse to aid identification. It was, after all, a pioneering field guide, showing clear signs of the fact that bird guide authors were still stumbling in the dark in trying to determine what information would best help observers to match a name with a bird. Besides, birdsong is notoriously difficult to render into words.

Still, Leach's single word seriously understates the exquisite beauty of a song often described as a 'falling-leaf melody'. Or, as a later field guide by Michael Morcombe puts it, the bird's song:

Usually begins with several loud, piercing high notes immediately followed by pure, high, clearly whistled, violinlike notes that descend in an undulating, silvery, sweet cascade, at times lifting briefly, only to resume the downward, tumbling momentum. Abruptly returns to the initial louder, sharper notes to repeat the whole sequence, often with slight variations.[2]

From 'Musician' to that extended encomium testifies to the increasing amount of information packed into field guides.

The modern field guide also names the species differently. Discarding the dreadful moniker 'flyeater', it is now called the White-throated Gerygone. Doubtless, the bird eats flies, but that's no reason to give it such an ugly name. It's one of those birds whose generic name has become its vernacular name; in this instance, with exceptional aptness, for the word 'gerygone', bequeathed by the 19th-century ornithologist John Gould, alludes to the bird's most appealing feature. 'Gerygone' has been variously translated (from Greek) as 'child of song', 'born of sound', and (my preferred rendition) 'born of song'. Whichever way, it's a fitting title, which is why Gould awarded it, first to the white-throated member of the genus. As a generic name, it's shared with other species, ten Australian gerygones being currently recognised although one, the Lord Howe Island Gerygone, is extinct. None of the others, in my estimation, are as gifted songsters as the white-throated, but all are musically talented.

Between 'flyeater' and 'gerygone', the white-throated species took several other names, including 'White-throated Warbler' and 'Bush Canary'. Though preferable to 'flyeater', they lack the musical lilt of 'gerygone'. I don't advocate the widespread use of generic as vernacular names. Gould made a habit of it, with unhappy results such as Ground Graucalus, Little Chthonicola, and Jardine's Campephaga. Those, thankfully, have long since been dropped for, respectively, Ground Cuckoo-shrike, Speckled Warbler, and Common Cicadabird. For those birds, the generic names have changed anyway, indicative of the instability of scientific as well as vernacular names. In just a few instances, genus names have survived as vernacular names, usually when they have a pleasant ring. 'Sitella' is an example, although in that case the generic name has been retained as a vernacular long after the bird lost it in science. The Varied Sitella is now in the genus *Daphoenositta*.

In his 1970 book, *Australian Warblers*, Arnold McGill devoted some discussion to the choice of vernacular names, expressing a preference for those that were pleasant-sounding or aptly descriptive of plumage, call, or habitat. Sometimes, he suggested, genus names fitted the bill:

> The 26-lettered Golden-headed Fantail-Warbler is cumbersome and Golden-headed Cisticola, or even Golden Cisticola is recommended. In fact the generic name of warblers could be applied in some cases as the vernacular title also: not only Cisticola but also with Calamanthus, Hylacola, Gerygone, Malurus, Sericornis and Amytornis, all of which names "roll off the tongue" far better than the numerous hyphenated "wrens" and "warblers".[3]

Most of those names failed to take root as vernaculars. 'Cisticola' and 'gerygone' did, though not without resistance.

In May 1978, after decades of debate, the Royal Australasian Ornithologists' Union (RAOU) issued its 'Recommended English Names for Australian Birds', devised by a committee of six chaired by the eminent ornithologist Richard Schodde and including McGill among its members. Replacing 'warbler' with 'gerygone' was among its recommendations, for which it offered explanations and justifications. 'Gerygone is a simple word, easily pronounced, [which] could enrich the language', it maintained, adding that 'catch-all names such as "warbler"' were inappropriate for Australian birds 'because these names imply associations with groups in the Old World, which may be quite unrelated'.[4] However, the RAOU names committee offered no translation of 'gerygone', although that would have bolstered its case.

Howls of outrage followed. Disgruntled birders complained that the new names were elitist, imposed by self-styled experts, would alienate the public, and would inhibit the popular appreciation of birds. According to Rex Sharrock in the *Bird Observer*, names such as 'gerygone'

> are absolutely meaningless to all except those who have an extensive knowledge of our birds. The use of such names only widens the gap between ornithologists and those with a less expert knowledge, that is, interested members of the public, and certainly does nothing to encourage Bill and Mary to appreciate more the bird life around them.[5]

In a more light-hearted vein, Rex Buckingham contributed a piece of what he acknowledged to be 'ragged doggerel' titled '"A Birdo's Lament" or "More Names, Less Birds"'. Among its many lampoons of the RAOU's recommended names is the following:

> Where Warblers warbled fancy-free/ Now flits the strange Gerygone ...

> Fond birdoes squirm and cry 'Desist/ From changing name or changing list'.[6]

At least Buckingham's rhyme and rhythm show he knew how to pronounce the word. Despite claims that the name is easy to pronounce, some birders took a while to learn that 'gerygone' has four syllables and that the final 'e' is voiced.

However named, the White-throated Gerygone is not only a silvery songster, but also a beautiful bird. Rich yellow below, except for the white throat that gives the bird its name, and grey-brown above, it's a tiny sprite that can be difficult to see as it forages in the treetops or momentarily sits still to sing. It's more often heard than seen, and its invisibility can add to the exquisiteness of its song. A seemingly disembodied melody floating from somewhere ethereally above lends a special magic to a bird named for being 'born of song'.

Most other gerygones are plainer than the white-throated species, although the Fairy Gerygone of coastal Queensland is a handsome bird of predominantly yellow and olive plumage. It's not clear why this, rather than any other species of gerygone, earned the sobriquet 'fairy'. All are fairy-like in their diminutive, fluttery attractiveness; and as McGill remarked, they pour 'fairy-like melodies from the treetops'.[7]

Although I've awarded the best songster title to the White-throated species, not everyone agrees. McGill suggested the Large-billed was 'possibly the best songster of all the *Gerygone* species', quoting acclaimed zoologist Jock Marshall to back up his judgement.[8] Others have given the gong to the Western Gerygone. It is certainly an accomplished singer, with perhaps a bit more variety in its melody than the White-throated. An extraordinary description of its song was given by ornithologist Alan Bell, as quoted in Dom Serventy and Hubert Whittell's *Handbook of the Birds of Western Australia*: 'I have never heard sounds so plaintively microscopic, so clear and yet scarcely perceptible. The ghost of a kitten's mew—the echo of dwarf violins played in the moon—these were the bird's notes.'[9]

That's poetry rather than ornithology. Although Serventy was a professional ornithologist with a PhD in zoology—one of the few in Australia so credentialed in the mid-20th century—his appreciative quotation of Bell's paean clearly shows his love of both the birdsong and the poetry.

Birdsong has a universal appeal to humans. Appreciation of bird music spans cultures, places, and times, bearing out the contention of commonalities between the aesthetic sensibilities of birds and people. Why we share such tastes is uncertain, although in his popular ornithological study, *Where Song Began*, Australian ecologist Tim Low suggests that birdsong may have influenced 'the evolution of human acoustic perception, and in particular our sense of what sounds pleasing'. More forthrightly, Low states that, 'Birds have of course influenced human music', instancing compositions by Beethoven, Haydn, and Messiaen.[10]

He could have found instances closer to home. In 1964, the composer Nigel Butterley penned a piece inspired by birdsong he heard from the treetops near his home in the Sydney suburb of Beecroft. To ascertain the name of the bird, he phoned Alec Chisholm, introduced himself, and whistled down the line a song that Chisholm instantly recognised as that of the White-throated Gerygone. 'It's a novel business', Butterley remarked, 'for a musical composition to be "christened" through an imitation of a bird's call whistled over the telephone.' Bearing the bird's then-current name, 'The White-throated Warbler' was first performed in Sydney on 27 February 1965 by recorder virtuoso Carl Dolmetsch, accompanied on harpsichord by Joseph Saxby.[11]

Chisholm delighted in Butterley's work, both as a musical composition and as a tribute to the musical talents of Australia's birds. Promoting the latter had been among Chisholm's hobbyhorses since his first forays as a writer in the early years of the 20th century.

He and numerous others, ranging from museum ornithologist A.J. North to literary nationalist Nettie Palmer, earnestly rebutted the image of Australia as a land of 'songless bright birds', as the poet Adam Lindsay Gordon had described it in 1870. Confuting the 'songless birds' stigma became almost platitudinous among the growing cohort of Australians who, in the early 20th century, sought to anchor a sense of nationhood in the natural environment. Yet the fact that well into the century they felt a need to rebut it—and rebut it vehemently—suggests that the 'songless birds' image held some public credibility.[12]

Partly, it was the insecurity engendered by the 'songless birds' slur that prompted many Australian birders in the first half of the 20th century to compare Australian with overseas birdsongs. Such comparisons recur frequently through the Australian bird literature of those decades, and the comparisons were almost invariably with British birds.[13] Australia then was closely bound to Britain; Britishness was part of the Australian identity; and good-natured rivalry permeated the relationship (although occasionally rivalry soured into enmity, as in the infamous 'bodyline' cricket tours of 1932–33). Rivalry over birdsong never reached anywhere near the intensity of bodyline, and the competitiveness was entirely one-sided, insofar as British birders seldom, if ever, entered the fray by barracking for their birds against antipodean rivals. But for many Australian birders it was a point of national pride to assert the superiority of our avian songsters over those of Britain.

In the numerous comparisons between Australian and British songbirds made in the early decades of the 20th century, the White-throated Gerygone never scored a mention. At least, I've been unable to find a mention despite diligent searches. Australians at that time certainly appreciated the beauty of the gerygone's song, extolling it frequently and fulsomely. But for purposes of competitive comparison, the White-throated Gerygone's wistful strains seem not to have had hit the right notes. The usual birds that Australians championed against British rivals were robust carollers such as the Pied Butcherbird, Australian Magpie, and Grey Shrike-thrush, or loud and versatile vocalists such as the Superb Lyrebird, or birds with close British relatives and hence similar songs, such as the Reed Warbler.

An ardent nationalist, Chisholm staunchly upheld Australian birds' pre-eminence in song, a judgement reinforced during his visit to the UK in 1938.[14] Gerygones had a special place in his heart, but not in his international birdsong comparisons. Affectionately calling them 'elfin warblers', he went into rhapsodies over their songs, especially that of the white-throated species:

> Stand in a suitable forest on a bright day in spring and you may hear a fairy bird-voice float from a tree-top—a delicate, ethereal voice that is at once happy and wistful... Probably this little dryad, the white-throated warbler, is more certain of itself, and more tranquil, than its melody suggests to human minds; but the fact is that the wistful quaver and the dying fall promote an appeal more potent than that of many stronger and richer bird melodies.[15]

It's an evocative rendition of a lovely song, and, probably without intending to, Chisholm put his finger on what it was in that song that inhibited him and his birding compatriots from using it in antipodean musical contests with Britain.

While birders appreciated the aesthetics of birdsong, they were equally appreciative of avian vocalisations as means of identification. They always have been. Although Leach gave scant attention to calls in his 1911 field guide, that was due to the novelty of the genre, not to birders' inattention to calls and songs. It's clear from the writings of Leach's birding contemporaries that they relied on vocalisations to identify birds at least as much as birders do today—probably more, since their optical equipment was so rudimentary. Then, as now, proficiency at identifying by call was a point of pride among birders, especially those with exceptional mastery of the skill. On the Lamington Tableland in 1920, for example, ornithological collector Sid Jackson boasted of his ability to 'pick out an Atrichia [Rufous Scrub-bird] or Olive Thickhead [Whistler]—even if they only utter a single note—and a long way off'.[16]

In his 1945 field guide to the birds of Tasmania, Michael Sharland asserted that:

> Too much emphasis cannot be placed on the value of the bird's note or song as a guide to identification. If the observer memorises these notes half the trouble is over. It is not exaggerating to say that practised observers rely more on the note or song as a means of determining a bird's identity than on any other single characteristic or feature in the field.

Excellent advice, but Sharland immediately went on to explain that while he tried 'to set down phonetically the call notes of bush birds', the difficulties were so great that accurate and helpful renditions of bird calls were often 'not able to be conveyed in print'.[17]

One way of representing bird calls is to render them into words that supposedly resemble the call. The Willie Wagtail's 'sweet pretty creature' is an example familiar to many Australians. But resemblances often seem stretched or idiosyncratic. Edward Sorenson, who wrote charming stories about birds for early-20th-century Australians, gave these renditions:

There is an interesting bird found in the scrubby regions of Southern Queensland and Northern New South Wales. The female calls "Hope you're well!" The male, being less modest, answers "Go to—!" The catbird cries, infant like, "Where are you, Maria?" and a little-weather prophet, influenced by atmospheric conditions, shrills forth "Twill-I-I wet you!" The willy wagtail, executing a graceful pirouette on a cow's back, invites you to "Get y'r pitcher!" Another bird, with a deep voice, demands "Whaffor!" while the mopoke makes perennial request for "more pork."[18]

The last is conventional, although the others may puzzle birders today. However, such transliterations into speech, sometimes called 'warblish', were not aimed purely at identification, but also—perhaps primarily—at fostering feelings for birds. As one exponent explained in 1925, 'Syllabising bird notes is a happy fancy of childhood and of all bird lovers, and there is no doubt that the little feathered creatures are more endeared to young and old when they are known as speaking birds.'[19]

At the more sophisticated end of the spectrum, birdsong can be represented in musical notation. In 1916, Robert Hall published an article on the 'Morning Song of the Noisy Miner' in the *Emu*, with a full-page plate setting down the song in musical notation with accompanying syllabic approximations. Hall was a museum curator and ornithological collector, one of the more rigorously scientific members of the RAOU that he had helped to found. Nonetheless, he clearly appreciated the 'pure joy' of the miner's morning song—'a psalm of dawn'—and transcribed the song not only as an empirical record, but also as a tribute to the bird's musical talents.[20] Musical notation is a comprehensive way of rendering birdsong, but it is too complicated to be useful for identification, and too specialised to boost the popular appreciation of birds.

Similar pitfalls accompany the sonograms that have been used to depict bird vocalisations. Sonograms diagrammatically graph the frequency of a sound (on the vertical axis) against time (on the horizontal). They appear in such specialist works as the *Handbook of Australian, New Zealand and Antarctic Birds*, and have been used in some American field guides, but are too technically demanding to have established a foothold in many books intended for the general public. As the American birding historian Scott Weidensaul wryly observes, sonograms are 'visual representations of bird sounds that only someone with a Ph.D. in acoustics could interpret'.[21] No major Australian field guide adopted sonograms.

Most have rendered bird vocalisations through verbal descriptions and/or syllabic approximations, as in the following examples for the White-throated Gerygone's song taken from a cross section of field guides over time. Neville Cayley's *What Bird Is That?* (1931) said that this bird's 'spring song, a sweet cadence uttered at frequent intervals throughout the day, is one of the finest of our bird melodies'.[22] That's laudatory, but offers no help in identifying the bird. Most of Cayley's descriptions of birdcalls are like that. Peter Slater, in his path-breaking field guide of 1974, described this species' song as a 'liquid descending trill "wh-wh-whee-hoo-whee-hoo whee-hoo whee hoo whee hoo whee hoo ... whee-youuuuu"'.[23] A later guide by the three Slaters—Peter, Pat, and Raoul—gives a better rendition as a 'Lovely descending melody, evocative of falling leaf, followed by more explosive "phee-ee-ew"'.[24] Pizzey and Knight's popular field guide also follows convention in describing the gerygone's voice as an 'oft-repeated beautiful, silvery "falling-leaf" of song in minor key, with upward recovery toward end, tailing off'.[25] Somewhat similarly, a recent CSIRO guide renders the White-throated Gerygone's voice as a 'Sweet-sounding, far-carrying, musical song: initial quick whistling notes blend into a long liquid cascading trill, rippling in pitch but becoming lower'.[26] Typical of field guide descriptions, these characterisations of the gerygone's song interlace aesthetic with empirical elements.

Going several steps further, a new generation of field guides lets us hear the gerygone sing in all its aural glory. In 2014, Michael Morcombe's *Birds of Australia* became the first Australian field guide to be transformed into an app, with recordings of thousands of bird vocalisations among its suite of identification cues. Several more such apps have since appeared. Granting unprecedentedly easy access to replayable replicas of birdcalls, they facilitate not only identification, but also the practice of calling birds in. It's not a new practice, having been done in earlier times with cassette tapes; before that, reel-to-reel tapes; and forever by mimicking birdcalls. But digital recordings on mobile devices make calling in the birds ridiculously easy, tempting some birders into overusing it and impelling others to deliberate on the ethics of playback. In birding, as in other human activities, technological advances prompt ethical questions.

The ethical questioning of audio playback arose only recently, but, as we'll see in subsequent chapters, the ethical dimensions of our interactions with birds have been salient issues throughout the history of birding. Few have loomed as large as the ethics of collecting, a practice once almost ubiquitous among birders. To that practice, I now turn.

CHAPTER THREE

Tooth-billed Bowerbird: Collecting

FOR MUCH OF THE year, Tooth-billed Bowerbirds are hard to find. Foraging in the canopy of upland rainforests in Queensland's Wet Tropics, the birds' brown backs and streaked underparts make them difficult to see in the dim light of their leafy, lofty homes. But come the mating season, from September to December, and they seem to be everywhere. It's then they build their distinctive display courts, unlike the bowers of other bowerbirds. Instead of constructing towers or avenues of sticks, the Tooth-billed Bowerbird carefully clears a circle of bare ground on the rainforest floor, usually between one and three metres in diameter and incorporating at least one tree trunk. On that bare patch, the male meticulously—perhaps obsessively—arranges freshly picked green leaves with the pale underside uppermost. Above this leafy display, he sits on a song-perch and pours out a remarkable, sometimes ear-splitting, concatenation of song, babbling, and screeching.

It's the display call that usually gives the bird away. The Tooth-bill is typically heard before it's seen, for although in the mating season it comes down from the canopy into the lower storeys of the forest, the light is still dim, and the bird's plumage cryptic. But the calls can't be missed. Displaying males intersperse their own songs with mimicry of Bower's Shrike-thrush, Bridled Honeyeater, Mountain Thornbill, and other avian songsters, sometimes rendered loudly and harshly, other times as a soft and melodious sub-song. In the mating season, the males seem never to stop calling, and the rainforest rings with their revelries.

It was by their calls that, in October 1908, ornithological collector Sid Jackson first located a Tooth-billed Bowerbird near Lake Eacham on the Atherton Tablelands. He went there to study this species and collect its nest and eggs, which hitherto were unknown to science. Hearing a medley of harsh, throaty notes and mimicked songs he thought likely to come from his target species, he identified it by a means no birder is likely to adopt today. In 1908, there were no field guides to Australian birds, so Jackson resorted to recollections of more tangible identification aids. 'At last I sighted my quarry', he recounted, 'and his thrush-like breast was turned towards me, so that from my memory of preserved specimens, which I had recently examined in the Queensland Museum, I recognized the lonely vocalist as indeed a veritable Tooth-bill.'[1]

Sid Jackson was among the most renowned field ornithologists and collectors in early-20th-century Australia. His specialisation was oology—the collection and study of eggs—but he also collected bird skins, and, like other collectors of his day, his interests were omnivorous, extending to terrestrial molluscs, sundry other animals, plants, and occasionally Aboriginal artefacts. In 1907, after decades of successful collecting as an amateur oologist, he was employed as curator and field worker for the wealthy private collector H.L. White of Belltrees near Scone, New South Wales. Jackson's north Queensland expedition of the following year was his first major collecting assignment for White and his first as a fully- fledged professional collector.

Jackson was an extraordinarily skilled tree-climber—a talent obviously advantageous for an egg collector. He was immensely proud, even boastful, of his skill, and posed for numerous photographs of himself aloft in the treetops. His arboreal agility is all the more remarkable given his portly figure—he weighed 16 stone (110kg)—and his contemporaries marvelled at the disjuncture between appearance and aptitude. Birders today are more likely to be amazed that he called himself a 'bird-lover' even though he shot birds and robbed their nests for a living.

However, in the early 20th century, loving birds and collecting them were not considered incongruous. A majority of members of the Australasian Ornithologists' Union (AOU; after 1910 Royal Australasian Ornithologists' Union, RAOU), founded in 1901, and the South Australian Ornithological Association (SAOA), founded in 1899, were collectors of skins and/or eggs. This they found quite compatible with their personal admiration for birds and their organisations' commitment to bird protection. So did their counterparts in Britain and North America.[2]

The first two decades of the 20th century were a heyday for ornithological collecting in Australia. Supplementing the already established museums, both the (R)AOU and the SAOA provided ornithologically focussed institutional bases for the men and the few women who collected and studied skins and eggs as part of their endeavours to advance the scientific understanding of Australia's avifauna. Disquiet over collecting was voiced, but it was not until after the First World War that disquiet escalated into acrimonious dispute over the rights and wrongs of shooting birds and robbing their nests. Until then, Australia's oologists and ornithologists could conduct their collecting forays with few pangs of conscience, confident that their activity was advancing the frontiers of knowledge of the avifauna of the continent. It was indeed, for, as the director of the Australian National Wildlife Collection, Leo Joseph, has remarked, the 'collections of yesteryear' are still of inestimable scientific and conservation value today.[3]

Jackson had few qualms about collecting nests and eggs. He seems to have assumed that the birds would build another nest and lay more eggs; probably, most did. Killing birds was another matter. Jackson did so, for both specimens and field identification, but dealing death evoked more complex emotions than stealing eggs.

Jackson's 1910 expedition to the Dorrigo scrubs in northern New South Wales culminated in his taking a specimen of the rare Rufous Scrub-bird. His target was the female scrub-bird, but, failing to procure the only one he encountered, it became his 'painful duty to shoot the male' to authenticate the nest he had collected. By the time he did so, he had been closely observing it for two months and had become emotionally attached to it. Jackson's diary records his deed and the feelings it evoked:

> I took careful aim just as he had finished imitating the notes of the [Logrunner], and I plucked up and shot at my dear old pet Atricia (although much against my will to do so) ... I then left the locality, and the shooting of this male bird has rather cast a gloom over me as it has silenced all further investigations, and it seems so hard to do so after the value and interest the poor creature has been to me.[4]

Regretful though he was, he acknowledged that the bird he shot was a member of a species under imminent threat of local extinction.

On his 1908 expedition to the Atherton Tablelands, Jackson shot comparatively few birds since his primary objective was to collect the nest and eggs of the Tooth-billed Bowerbird. This proved unexpectedly difficult. The birds were so common and their courts so numerous that after two weeks in the district he had found 112 Tooth-billed Bowerbirds' 'playgrounds', but, 'strange to say have not yet found a nest after all the great patience and most careful hunting, day after day'.[5] Jackson's 'hunting' was helped by others, including local timber-getter and amateur egg collector Ted Frizelle, with whom he camped on the banks of the Barron River near Tolga. Halfway through the expedition, they moved camp from the southern to the northern side of the river, to a site they called Cherra-chelbo after the local Aboriginal name for the Tooth-billed Bowerbird.

Jackson and Frizelle knew the name for the bowerbird because they had close interactions with the local Yidinji people.[6] In earlier

Figure 1. Sid Jackson (seated, centre) with Aboriginal assistants in their Cherra-chelbo camp, 1908.

expeditions in northern New South Wales, Jackson had employed Aboriginal people as collectors, as was then common practice.[7] He acknowledged the skills and contributions of his Aboriginal assistants; he singled out one man, Nymboi Jack from the Clarence River, for special praise; and he seems to have had amicable relations with local Aboriginal people.[8] On the Atherton Tablelands, relations were more fraught, and his attitudes toward Aboriginal people an uneasy composite of fear, apprehension, and admiration. Fear and apprehension are understandable, for the frontier wars between Tablelands Aboriginal groups and invading settlers were well within living memory, having ended only a few decades earlier. When Jackson arrived at Atherton in 1908, open warfare was over, but the local Aboriginal people were far from fully subjugated. Stumbling across an Aboriginal campsite in the forest, he 'judged it wise ... to give these wild warriors a wide berth'.[9]

But he did not give all Yidinji people a wide berth. He employed some as collectors, although apparently only two or three at any one time. One of them, named 'Mitchell' in Jackson's diary—presumably because he did not know this man's Yidinji names—earned his special admiration. On his climbing skills, Jackson could only resort to superlatives, describing Mitchell as 'truly a wonderful tree climber', 'a splendid climber', 'a capital climber', and 'a beautiful climber (champion of all I have met)'.[10] For Jackson, this was no small matter. He was intensely proud of his own climbing skills, and by acknowledging Mitchell's prowess in that domain he was offering a supreme compliment. He also praised Mitchell for his intelligence, diligence, trustworthiness, practicality, and knowledge of nature. Yet his respect for the man was offset by patronising assumptions of superiority.

Mitchell and Frizelle collected the expedition's first Tooth-billed Bowerbird's nest while Jackson was temporally absent at

Figure 2. Two of Jackson's Aboriginal assistants: on the left is the Yidinji man known to Jackson as 'Mitchell'; the man on the right may have been the one whom he called 'Billy'.

Evelyn and Atherton. When he returned on 5 December, he was sick with fever and in the depths of despair after weeks of fruitless searching. Even the nest did not raise his spirits, for it contained not an egg but a downy chick: 'a fearful disappointment', Jackson bemoaned. Nonetheless, he quickly rallied, and on 7 December he and Frizelle planned a raid on a Tooth-bill's nest they had been keeping under surveillance. For days, Jackson had been debilitated by fever, but now, anticipating a 75-foot climb into the canopy and taking the eggs 'by my own hand', he went to bed 'in a great fever of excitement'. He had a restless night, for 'all I could keep in my mind was Tooth-billed Bower Birds and their nests and eggs'.[11] Jackson invested powerful emotions in his collecting.

'Success at last', Jackson reported on 8 December. Mitchell and he climbed 80 feet into a tree they knew held a Tooth-bill's nest, and when they reached it:

> I slowly lift my head, and at last!—yes, at last!—my eyes actually rest upon the frail stick nest, which contains two lovely very dark cream-coloured eggs. I can scarcely realize the situation, my excitement being so great. I am trembling like a leaf from head to foot. That which has haunted me day and night—the principal object of my mission to North Queensland—has been at length discovered.[12]

According to Jackson, Mitchell was in 'high delight' at their achievement, and Frizelle, who had remained on the ground, was 'in a great state of excitement'. 'We left the scrub happy happy men', he told his diary.[13]

But they'd been beaten to the punch. Working in the Evelyn scrubs south of Herberton, another collecting team headed by George and John Sharp had acquired a Tooth-billed Bowerbird nest and eggs on 7 November. George Sharp was reputed to have engaged 60 Aboriginal helpers, as against Jackson's three. Perhaps that was why Sharp was so successful.

Having found a clutch of eggs on 8 December, Jackson and his team went on to collect as many as they could, following normal collectors' practice. The next day, they went to a tree they knew contained another Tooth-bill's nest. Mitchell climbed 90 feet to the nest, and on signalling that it held two eggs, Jackson and Frizelle 'jumped about with excitement' at the realisation that 'another set of these rare eggs should fall to us again'.[14] Frizelle was a brawny timber-cutter, so the sight of him and the portly Jackson jumping about in excitement at finding a nest full of eggs may have provided some entertainment for their Yidinji companions.

Jackson liked to emphasise the difficulties and drama of what he did. In one diary entry explaining the labours of finding Tooth-bill's nests, he declared:

> They are extremely shy birds, and they fly so fast and usually in a straight line through the dense scrub that a man has NO chance in the world of following them, and to find a nest is a great conquest and is the result of much patience and perseverance, climbing and examining trees and watching the birds day after day and locating them to a certain area of the scrub. That is the ONLY way to find them.[15]

He liked, too, to present himself as a man who faced the perils of his enterprise with pluck and fortitude. Collecting was doubtless difficult and dangerous, but there were other sides to his interactions with birds that were not so different from what birders do today.

It is clear from his field diaries and notebooks that Jackson did a lot of incidental birdwatching, not only during his Atherton Tablelands sojourn, but on all his many expeditions. Indeed, birdwatching for fun was his favourite pastime. In that, he resembles many professional ornithologists today, who, when workaday bird studies are over, don binoculars, and venture out to watch birds for pleasure. While traveling by train or coach, Jackson compiled lists of the birds seen and heard *en route*, apparently challenging himself to see as many as he could. These records were not part of his professional assignments. They reveal Jackson as an enthusiastic—perhaps compulsive—list-maker, rather like many birders today.

Throughout his north Queensland diaries, Jackson recorded his joy at being among the birds and hearing them sing. His entry for 25 December at Tolga is typically lyrical:

> Birds sang and whistled everywhere just as if they tried to greet me on the morning of Xmas. Before breakfast I walked along the edge of the scrub and the first loud notes which greeted me were those of the Black-headed Log Runner as it cried—"Chowchilla chow-chow Chowchilla chow-chow".
>
> Other birds which I heard plainly were—the Tooth-billed Bower Bird, Blue-bellied Lorikeet, Black-faced Flycatcher, Swainson's Graucalus, Lesser Pitta, Ashy-fronted Robin, Coachwhip Bird, Northern Oriole, YB Fig Bird, Ptilorhis Victoriae, Spotted Cat Bird, Pigeons, YT Sericornis, Y-spotted Honey Eater, Bower Thrush, Koels, etc, while in the forest on the hill near Tolga, the Pale Flycatchers were singing sweetly in the cool balmy breezes. This morning was a perfect Xmas morning for the ornithologist, and here I was in the midst of all the glorious bird life.

Later that day, while walking along the Tolga-Atherton road, he happened upon a Tooth-billed Bowerbird that appeared to be nesting. He instantly switched into collecting mode, scrutinised the forest for the nest, climbed 50 feet into the canopy when he located it, and assessed the right time for a raid.[16] The juxtaposition of enthusiasm for nest robbing and delight in birdlife can be jarring. It recurs continually throughout Jackson's diaries.

While Jackson appreciated the common birds around him, he became excitedly agitated when he encountered species that were rare or new to him. Whenever opportunity offered, he eagerly avowed the rarity or uniqueness of birds he saw and eggs he collected, in ways that bear comparison with the rarity-hunting of modern twitchers. Indeed, the comparison can be pushed further, for, as birding historian Stephen Moss has pointed out, twitching and collecting have in common competitiveness as a major motivation.[17] Beyond that, the twitchers' lists and photographs are analogous to the egg and skin collections of birders of earlier times: tangible mementoes of avian encounters that matter to the practitioner.

Early in his northern travels, at Hawkins Creek near Ingham, Jackson found fresh Cassowary footprints, and 'became quite delighted to realize that I was for the first time actually in the true haunts of this interesting bird'. He searched likely places 'in hopes of finding a set of their beautiful green eggs, but I had no luck. Oh what a grand find such a thing would be to me. It makes one dream of such finds, it does so with me repeatedly'.[18] The parallel with modern twitchers' excitement at seeing—and anticipating seeing—a new species is palpable, although Jackson was more enthused by the prospect of getting his hands on the beautiful green eggs than of seeing the shaggy black bird itself.

He loved getting his hands on skins, too, especially those of rare and beautiful birds. When Jackson first met fellow collector George Sharp in the Evelyn scrubs, the latter handed him 'several beautiful skins of the handsome male of Newton's Golden Bower Bird and also one female. These I handled in the flesh for the first time in my life, and not many persons have done so'.[19] Jackson's writings often express an appreciation of nature in a spirit of innocent and wide-eyed wonderment. Yet he also revelled in the fact that what he did and saw was out of the ordinary.

On Frizelle showing him, for the first time, two Tooth-billed Bowerbirds' nests high in the rainforest canopy, Jackson excitedly recounted how rare and difficult to see they were, enthusing over the fact that he was seeing something few others had the privilege of seeing:

> Both [nests] are most difficult to see, in fact nest No.2 could not be found if a man examined the tree from the ground for 20 years. They are MOST difficult to see and at the same time very small, and only consisting of a few sticks makes them all the less conspicuous of course. They will always be rare I am quite sure of that.[20]

Enthusiastic about rarities, Jackson's excitement peaked when physically performing his collecting exploits. There was the exhilaration of climbing trees to dizzying heights, the spice of danger, the edge of unpredictability, and, if successful, the palpable reward of lustrous eggs and delicate nests.

Even when climbing and egg-taking were deputised to others—usually Mitchell, sometimes another Yidinji assistant or Frizelle—Jackson became highly excited at the moment of collection. On 11 December, Mitchell showed him a Victoria's Riflebird nest he had found just behind their camp. 'Now the excitement was intense', Jackson enthused in one of his favourite phrases. To raid the nest, which was located in foliage that could not bear a man's weight, they erected a makeshift pole, secured with equally makeshift vines, up which Mitchell climbed to find two eggs. 'Oh the joy that followed', Jackson gushed. 'Mr E.D. Frizelle and I were on pins and needles until Mitchell got on the ground again, then the bag of treasures was opened and needless to say delighted us all.'[21] After the eggs were taken, the climbing pole was left in place so the nest could later be cut down, and nest and eggs reunited for a photograph. That was Jackson's usual practice for taking photographs of nests and eggs.

Jackson was among Australia's leading pioneer bird photographers, but behind his photos lay quite spectacular levels of interference with birds and their habitats. This was the case for other early bird photographers, too, in large part because of the technological limitations of the cameras of the day. But more than technology was involved. It was also a matter of values and attitudes.

In later times, photography was proffered as an alternative to collecting, but in the late-19th and early-20th centuries, relations between photography and collecting were mutually supportive. Almost all the early bird photographers were also collectors of skins or eggs, and they regarded photographs as supplements to, rather than replacements for, those tangible objects. Jackson was explicit about this, explaining that, while in the field, 'I always carried a camera with me when practicable, and succeeded in supplementing my oological trophies with many unique and interesting photographs.'[22] For him and his contemporaries, the photograph was a convenient image of an object (egg or bird specimen), but primary value rested in that object. It would be some decades before possessing the image came to be valued over possessing the object itself.

CHAPTER FOUR

Crested Tern: Photography

THE CRESTED TERN IS one of the commonest seabirds in Australia. Inhabiting coastal and estuarine waters all around the continent, as well as in Tasmania and other offshore islands, it usually stays close inshore. It seldom ventures inland, unlike another common seabird, the Silver Gull, familiarly known to Australians as the seagull. Unlike the seagull, too, the Crested Tern does not gather around garbage dumps in search of food, nor does it pester picnickers for chips and sandwich scraps. Instead, it hunts in more dignified, ternly fashion, flying over the sea and plunging down to seize small fish and squid.

Typical of its family, the Crested Tern has a subtle beauty, with plumage of pure white, delicate grey, and stark black, tipped off with a large straw-yellow bill. It can look a bit comical at times, its black crest resembling a hairstyle that varies, depending on wind and wetness, between a Beatles-style mop and punky spikes. It's a beautiful bird, nonetheless, and, like all terns, its flight is graceful, with deep, easy beats of its pointed, backswept wings. Mating flights are particularly spectacular. By way of preliminary, the male walks toward a prospective mate with a fish in his bill. If accepted, the two fly off together in balletic zig-zag flight for up to several hundred metres, and then glide back to land.

Courtship rituals complete, the female lays an egg (usually only one) in what passes for a nest in the tern's world. It's a mere scrape in sand or shingle, but what it lacks in engineering sophistication, Crested Terns make up for in sheer numbers—thousands nest together on islands, coral cays, and offshore sandbars. Often interspersed among other tern species, the nesting birds are tightly packed, each just beyond pecking range of its neighbours.

It was in such a situation that the Crested Tern attained historical distinction: it was the first Australian bird species to be photographed in the wild. A.J. Campbell took the photograph on Direction Islet, off Rottnest Island, Western Australia, on 21 November 1889. His photo shows a flock of Crested Terns, some

Figure 3. Australia's first bird photograph: A.J. Campbell's 'Rookery of Crested Terns by ocean', 1889.

airborne, but most sitting on their nests among patches of samphire or some similar plant. It was among the earliest photographs of wild birds taken anywhere in the world.[1]

But historical firsts are always contestable. A.J. Campbell's grandson, Ian Campbell, has disputed the status of his grandfather's Crested Tern photograph as the first of an Australian wild bird. According to Ian, the honour rightly belongs to another photograph by his grandfather, of an Osprey's nest containing a chick, 'which preceded the Rottnest picture by a few weeks!'[2] A.J. Campbell himself stated that the 'exact date' of this photograph 'was the 5th November, 1889', whereas he specified a date of 21 November for his Crested Tern image.[3] So Ian Campbell was right—maybe. For while the 5 November photograph clearly shows a huge nest with a man standing next to it, the image of the young Osprey is so hazy as to be unrecognisable, and A.J. Campbell's own words indicate that his photographic subject was the nest, not the young bird in it.

For the tern photograph taken 16 days later, by contrast, the birds themselves were the primary subjects; and although the photo is monochrome, the image is sufficiently clear to identify the species. Campbell described the occasion:

> Here is a scene to gladden the heart of any naturalist. It fairly thrilled me with delight. Standing upon the apex of the rock there appears before us a congregation, consisting of scores of these large handsome sea-birds. Their silvery dresses, relieved by black caps and yellowish bills, are in agreeable contrast to the dull-coloured rocks and vegetation, with their liberal dressing of bird-lime. The sea rolls calmly below. The day is superbly fine; a gentle breeze is flicking the sky with cirrus-clouds—just a light a photographer loves—so here I plant the camera and chance an exposure of the bird colony.[4]

As this excerpt shows, Campbell had a fine aesthetic appreciation of birds in nature. The almost-final words—'chance an exposure'—also point to the vagaries of bird photography at the time, for birders then could invest hours, even days, of effort into capturing an image, and as often as not something would go awry.

Before his historic photo of Crested Terns, Campbell had photographed nests and eggs. Perhaps his first such photograph—the first ever taken in Australia—was of the nest and eggs of a White-breasted Robin, taken in the Torbay district, Western Australia, on 31 September 1889.[5] His Osprey nest photograph on 5 November seems to have been in that tradition, rather than an attempt to capture a bird on film (actually, on glass plate, since that was the photographic technology of the day). Indeed, the reproductive phase of birdlife was Campbell's primary interest, for, as well as being a pioneering photographer, he was a collector of nests and eggs, and it was for his oological investigations and publications, including his two-volume *magnum opus* of 1900, *Nests and Eggs of Australian Birds*, that he was best known in his own lifetime.

Nests and Eggs was illustrated with photographs by Campbell and his birding friends Sid Jackson and W.H.D. Le Souëf. Only a few photographs were of birds, among them that of the Crested Terns taken by Campbell on 21 November 1889. Most photos were of nests and eggs, a preponderance determined not only by the topic advertised in the book's title, but also by the technological limitations of the cameras of the day. Even when birds were the photographic subjects, those technological limitations ensured the photos were of birds at nest. A few photos illustrated the activity of egg-collecting, including a wonderful shot of a naked man climbing out of a swamp and up a tree to collect a Wood Duck's nest. Perhaps unintentionally, it captures the flavour of robust, masculine adventure so characteristic of ornithological collecting. Few egg collectors were women.

Figure 4. 'Taking a Wood-Duck's nest' by A.J. Campbell, 1894.

For more than half a century after Campbell's book, birds at nest remained the staple for bird photographers, because the nest was where a bird could be expected to sit still for more than a moment. Bird photography would soon dissociate from egg-collecting, and photographers would proffer their pastime as an ethically superior alternative to oology. But bird photographers were as powerfully drawn to nests as their egging alter-egos had been. The image rather than the egg became the prize, but for both collectors and photographers the nest was the focal point of their pursuits. Like egg collectors, too, early photographers risked life and limb in quest of nests, so that for many decades heroic tree-climbing was as characteristic of bird photography as it was of egg-collecting.

Egg collectors didn't just collect eggs. Like all birders, then and now, they also watched birds, admired them, studied them, and in myriad other ways appreciated the creators of what they collected. One of the wonderful things about Campbell's *Nests and Eggs* is that much of its text is not about nests and eggs at all, but rather celebrates birds in all their glory. Campbell described birds' appearance and habits in detail, and delighted in the beauty of birds, lauding, for example, the 'gorgeously-plumaged' Paradise Riflebird and the 'handsome Fishing Eagle [Brahminy Kite] in snow-white and rich chestnut plumage'.[6] Expressing a wide-ranging fascination and admiration for birds, Campbell's collections of nests and eggs might be regarded as the material manifestations of his interest and appreciation. That's an accurate but not sufficient explanation for his and his contemporaries' oological activities. Beyond that, they believed that, by collecting, comparing, and cataloguing eggs, they were contributing to the advancement of science. They were right.[7]

For a birder today, reading Campbell's *Nests and Eggs* can be a jarring experience. You'll find delightful passages glorying in the gorgeousness of birds, enthusing over their quirks of behaviour, telling of the tribulations endured to see and photograph them, and discussing the perennial ornithological issue of whether or not particular varieties were separate species. Then, suddenly, you'll bump up against passages not just narrating, but celebrating how Campbell shot birds and robbed their nests. Similar juxtapositions were common in most birding works published around that time.

Campbell collected eggs not only for his oological studies, but also to eat. His *Nests and Eggs* frequently and frankly acknowledges the fact. 'When fresh, the eggs [of the Mallee Fowl] are excellent eating; to this I bear testimony, having had one fried for breakfast. It was exceedingly palatable, being rich and delicate.'

'Contrary to expectations', he wrote of Emu eggs, 'the flavour was, if anything, milder than that of the domestic fowl's ...; therefore some palates may consider the Emu's egg tasteless, but we proved it a delicacy.' His account of Muttonbirds (Short-tailed Petrels in Campbell's day; Short-tailed Shearwaters in ours) devotes a great many words to the then-popular pastime of egging, including narratives of two of his own Muttonbird egging excursions to Phillip Island in 1884 and 1897.[8] Years later, as the incoming president of the AOU in 1908, he led the organisation's annual campout, which that year was held on the same island. He and fellow AOU members, along with their spouses and children, egged with gusto. '*Puffinus* eggs and bacon (fried) make a most *recherche* meal', Campbell declared.[9]

Just as he collected eggs for consumption as well as for science, Campbell also shot birds for both purposes. His accounts of hunting and eating birds, alongside his admiration for them, can be particularly jarring for today's readers. In *Nests and Eggs*, he marvelled at 'how majestic the Bustard looks as it paces slowly, with head erect'. Four sentences later, we're told that its flesh 'is well flavoured and excellent eating', followed by advice on how to shoot bustards, how to hang them in preparation for the pot, and an anecdote about one of his own bustard-shooting exploits. In another passage in *Nests and Eggs*, he commended the table quality of Brolga: 'After being hung for a few days and then stuffed with good gravy-beef they are excellent eating.' In similar vein, he averred that 'the rich flesh of a young Mutton Bird ... is decidedly delicate and delicious in flavour, if properly cooked'.[10]

Campbell was not unusual among late-19th- and early-20th-century birders in also being a hunter, nor in remarking on birds' table quality in their ornithological studies. Charles Belcher prefaced his 1914 publication, *The Birds of the District of Geelong*, with the hope 'that my little book may be the means of communicating to the general reader something of the enduring charm and delight which from childhood I have found in the observation of wild birds'.[11] He was entranced by birds, and rejoiced in watching them, but frankly acknowledged—and celebrated—his exploits as a hunter, and commended the table qualities of a wide range of species, from Freckled Duck to Sharp-tailed Sandpiper. Perhaps it's unsurprising that birders who shot for the specimen drawer also shot for sport and the table.

Indeed, the line between hunting birds for specimens and hunting them for sport could be thin. Throughout Campbell's prolific writings on his collecting adventures, his admiration and feeling for the birds he shot shines through. So does the thrill of the chase and the excitement of the kill, as in the culminating moment in his account of stalking a Victoria's Riflebird on the Barnard Islands:

> In an instant, reckoning on the intervening obstruction, I discharged No. 6 in lieu of dust shot. I was immediately surrounded by thick smoke hanging in the damp air, but whether my beautiful feathered visitor had fallen or flown I knew not. Overcome with excitement, I felt as if I could hardly venture to ascertain. I crawled slowly up the gully through prickly creepers, and on parting a bush there I beheld a gorgeous cock rifle-bird dead upon its back. It was a beautiful object in its rich shining garb.

At the end of the day's collecting on the Barnard Islands, Campbell tallied up the kills: 'grand total, 17 birds of paradise—the greatest day's taking of rarities recorded in the annals of Australian ornithology. Certainly it was a most unfortunate day for the poor birds, but for their sake let us hope it may never occur again.'[12] Like other collectors, he delighted in securing specimens, especially of rare and beautiful birds, while at the same time regretting the blood spilled in the process. Killing was not just a duty he performed for science, but an achievement in which he took pride, tempered with qualms of conscience.

As the 20th century progressed, it became less common for birders to shoot birds for sport, as it also became less acceptable to shoot them for specimens. In line with changing public opinion, those birders who did hunt became increasingly reluctant to acknowledge the fact. But some did. Jack Hyett, for example, was a renowned amateur ornithologist, editor of the *Emu* from 1965 to 1968, and recipient of the 1985 Australian Natural History Medallion. He was also a keen hunter, and was perfectly frank about it in his amiable writings on natural history.[13] So was the wonderfully named Brigadier Hugh Officer, who, in his 1978 *Recollections of a Birdwatcher* recognised that, 'Some readers may be surprised to hear a bird watcher talking about shooting birds. I enjoy game bird shooting and I am not ashamed to admit it.'[14] In the 1960s and 1970s, Brigadier Officer was among Australia's most acclaimed birdwatchers, renowned for his infectious enthusiasm for seeing new species.

Hunting, in the sense of shooting to kill, has disappeared from birding practice, other than in the highly regulated domain of scientists securing specimens for study. Even there it is controversial, and some birders protest vehemently against specimen-taking for scientific research. But hunting—at least sports hunting—has always been more about the chase than the kill. And in modern birding practice, the pursuit of the bird, be it a rarity or common species, remains central to the pastime. How birders pursue their quarry is not so different from how hunters do it, except for omitting the culminating moment of bloodshed.

Birdwatching is a form of sublimated hunting in which the skills of locating, stalking, and identifying birds are highly prized—indeed, they're fundamental to the pastime. Instead of a kill, it culminates in a tick, a note, a photograph, or even a memory: those are the mementos of a bloodless form of hunting, but one that involves the thrills and skills of the chase associated with its more brutal counterpart. At least some birders acknowledge the continuities. In their 1998 how-to book on *Birdwatching in Australia and New Zealand*, for example, Ken Simpson and Zoë Wilson note that, 'Birdwatching is a kind of hunting, except that our "prey" must never be harmed.'[15] True enough when they wrote those words, although 90 years earlier the saving proviso was usually omitted.

Like birdwatching, bird photography retains strong parallels with hunting, exemplified in the language of 'shooting' with a camera. Bird photographers have often been more explicit than that. R.T. Littlejohns, who in the early 20th century implored his fellow birders to abandon the gun and take up the camera, drew explicit analogies between his preferred practice and hunting. 'Hunting Birds with a Camera' was the title of an article he contributed to a 1949 issue of the *Bird Lover*, a magazine for children. In it, he gave practical advice on how to photograph birds, and ethical advice on how not to harm them in the process. But he persisted with the analogy, concluding the article with the injunction 'GOOD HUNTING!'[16] A photographer of a later generation, Peter Slater, advising birders on how to sharpen their skills, characterised 'stalking birds with the camera' as 'the art of benign hunting'.[17]

Donald Trounson, who in the 1960s initiated the National Photographic Index of Australian Birds, represented his avocation as a modern version of a primal impulse. Seeking to 'popularise the attractions of bird photography as a hobby', he maintained that 'it is a pursuit which satisfies one of the most primitive and fundamental urges in the make-up of man: the urge to hunt, to pit his skill against the wits of some other living creature, and enable him to acquire his trophy without the least harm coming to the quarry'.[18] The concluding proviso had long been crucial to those bird photographers who asserted the moral superiority of their activity over collecting. Trounson also put his finger on an important point by referring to the 'trophy'. Like the hunter and the collector, the bird photographer takes away an enduring memento of her or his encounter with a bird, to an extent that no other form of birding can match. A physical likeness trumps a mere memory or pencilled tick. That partly accounts for the growing popularity of bird photography as birding grew away from its collecting past, and its hunting impulses were sublimated into a chase without a kill.

Did A.J. Campbell have any inkling that the mode of recording birdlife he initiated in Australia, with his 1889 photograph of Crested Terns, would transform the conduct of birding? Probably not at that time. He used photography as a supplement to, not a replacement for, collecting eggs and skins. But long before his death in 1929, he would have been aware of the turning tide. By the time he helped found the AOU at the turn of the 20th century, there were murmurings of unease, especially over the needlessly excessive amount of collecting by amateur birders.[19]

Overseas, the anti-collecting voice was louder. The British birder Edmund Selous, who is credited with coining the term 'bird watching' in a 1901 book of that title, explicitly offered watching as a civilised alternative to the 'monstrous and horrible' practice of collecting.[20]

A prominent Australian opponent of collecting, also a founder of the RAOU and a pioneering bird photographer, was Charles Barrett. He abandoned collecting around the time of the union's creation, later recalling that he 'gave up ornithology to become merely a bird-

observer, because I could not shoot birds without feeling a brute'.[21]
More and more birders came to the same view, and, like Barrett,
took up the camera to avoid feeling such qualms.

CHAPTER FIVE

Plumed Egret: Protecting

WHEN THE NEWLY MINTED Australasian Ornithologists' Union launched the *Emu* in October 1901, the journal carried the subtitle *A Quarterly Magazine to Popularise the Study and Protection of Native Birds.* Its very first article underlined the imperative of the second of those objectives, warning members that 'it must ever be remembered that ornithologists and bird-lovers will have to "hammer, hammer, hammer" at some very apathetic skulls before due protection is achieved. The public must first be roused, then never be permitted to ignore the desired result'.[1] It's a warning as relevant today as it was then, and over the intervening 120 years and more, birdwatchers have been prominent in 'hammer, hammer, hammering' on apathetic skulls to advance the conservationist cause. During that time, conservationist campaigning has changed in myriad ways, but has always woven together emotional, aesthetic, ethical, scientific, and pragmatic appeals for the preservation of birdlife.

The first issue of the *Emu* also carried an article on 'Bird Protection' by Frank Littler, who warned that if the 'ruthless slaughter' of our birds was allowed to continue, it was 'only a question of a few years and many of our birds will be as extinct as the Dodo'.[2] Birders in early-20th-century Australia were acutely aware of the appalling record of bird extinctions overseas, and of imminent bird extinctions in North America, where the Passenger Pigeon, Whooping Crane, and Carolina Parakeet hovered on the brink. They knew that, as a recently colonised land, Australia was in a situation akin to North America, and were eager to avert the avian extinctions looming on the horizon here. The spectre of extinction haunted early-20th-century birding. It still does today.

Littler's 1901 article carried emotional and aesthetic appeals to save 'our feathered friends ... with their gay plumages'. It advanced ethical arguments, asserting that birds had a 'right' to what they needed for subsistence, and asking:

> Are the fruits of the earth to be man's wholly and solely? I think not, more especially as man, in many instances, takes away from birds the opportunities of sustaining themselves on their natural food as was given them from the first.[3]

However, Littler devoted most of his words to pragmatic reasons for conservation, especially birds' role as destroyers of insect pests. Giving primacy to utilitarian arguments for bird protection, with aesthetic, ethical, and other considerations in subsidiary positions, typified birders' conservationist pleadings in the early 20th century. Titles alone bear out that weighting, in books such as Robert Hall's *The Useful Birds of Southern Australia* (1907) and Walter Froggatt's *Some Useful Australian Birds* (1921), although neither of those made usefulness the sole criterion for the worth of birds.

Some disputed the stress on usefulness. Charles Belcher, a founder member of both the RAOU and the Bird Observers' Club

(BOC), stated in 1914 that 'the argument that we should study and protect our native birds because of their economic utility leaves me rather cold'. Birds helped control pests, he acknowledged, but their true worth lay beyond their utility:

> I find the most compelling claim of Australian birds upon our affections to lie not so much in their money value as in the direct influence of beauty which they will exert upon anyone who cares to open his eyes and ears to the life that is all about him by green forest, open plain, or sounding shore.... No single bird is there but has some peculiar beauty.[4]

None of Belcher's fellow birders would have disputed his affirmations of the beauty of birds. And there are clear indications that many who gave primacy to utilitarian arguments did so out of expediency: arguments assigning an economic value to birdlife were most likely to resonate in the corridors of power. Nonetheless, it was usually utilitarian arguments that were pushed to the fore in the conservationist cause.

However, the biggest bird-conservation campaign in the early 20th century foregrounded ethical rather than utilitarian arguments. The campaign was against killing birds for their plumes, then extensively used in ladies' millinery. It was an international campaign against an international trade, but Australia played significant roles in both. The quantity of plumage offered for sale at the time is staggering: a contemporary critic estimated that the feathers of tens of thousands of birds passed through the London markets alone in a single day.[5] So was the diversity of birds that fed the millinery trade, from seagulls to birds of paradise, from wrens to emus. Sometimes, entire birds adorned hats; sometimes, bits of birds such as wings and heads; often, just feathers. Among the most prized of the latter were the breeding plumes of egrets.

The plumes, called 'aigrettes' and (confusingly) 'ospreys', came from virtually all the world's species of egret. In Australia, the preferred source was the Plumed Egret, although Great and Little Egrets were also shot for their plumes. The preferred time for shooting was the breeding season, when the birds were most vulnerable and their plumes were at their prime. This meant that for every egret pair killed, their nestlings would suffer an agonising death from neglect and starvation. The emotional impact of confronting plume-wearers with that suffering was not lost on opponents of the trade.

A picture of an egret plume captioned 'The White Badge of Cruelty' often featured in anti-plume-trade propaganda. Most campaign literature was addressed to women, typically combining ethical injunction with emotional appeal. A 1910 leaflet by the Wild Life Preservation Society of Australia, for example, positioned the following words above a photograph of dying egret chicks:

> Every woman who purchases or wears one of these plumes is as directly responsible for the tragedy depicted below as the gunner who fires the shot.
>
> Can feathers obtained in such a way make any women fairer?
>
> Can true womanhood sanction such barbarous cruelty?[6]

Campaigners in America and Britain confronted plume-wearers with similar language, making similar appeals to feminine sensitivities and humane values. Against the slaughter of egrets, passions were inflamed, and moral righteousness averred.

Australia's biggest contribution to the global anti-plume-trade campaign was a series of photographs by amateur ornithologist

Arthur Mattingley, shockingly exposing the carnage behind the fashion for feathers. In the summer of 1906, Mattingley revisited an egret breeding colony in the Riverina that he and a colleague, J.A. Ross, had earlier surveyed. Approaching the colony by boat, Mattingley encountered:

> a sight that made my blood fairly boil with indignation. There, strewn on the floating water-weed, and also on adjacent logs, were at least 50 carcasses of large White and smaller Plumed Egrets—nearly one third of the rookery, perhaps more—the birds having been shot off their nests containing young. What a holocaust! Plundered for their plumes. What a monument of human callousness! ... How could anyone but a cold-blooded, callous monster destroy in this wholesale manner such beautiful birds, the embodiment of all that is pure, graceful and good.

Those were Mattingley's words in the RAOU's ornithological journal, the *Emu*, which at that time freely published such emotive outbursts alongside dry dissertations on avian taxonomy and rollicking tales of birdwatching adventures. Mixing genres in a manner that a later generation of savants celebrated as 'postmodern' was commonplace in the early *Emu*.

Mattingley's own article on the slaughter of egrets had elements of birding adventure. He recounted at length how, to get good photographs of orphaned egret nestlings, he ascended their nesting tree using climbing irons, but when about 25 feet aloft he was stung by a hornet, paralysing his right hand and temporarily disabling him. At almost the same moment, his climbing irons slipped, so he fell into the swamp below. 'My clothes were damped, but not my ardour', he declared, 'and I managed to send up the rope ladder, and re-ascended the tree, where I secured another snap-shot of the poor starvelings from a precarious coign of vantage.'[7] Mattingley's photographs were hard won, but their impact reverberated around the globe.

Seven of his photographs were published in 1909 as a pamphlet, *The Story of the Egret*, by the London-based Royal Society for the Protection of Birds (RSPB), which had taken the lead in campaigning against the plume trade in the United Kingdom. Graphically depicting the birds' drawn-out deaths, the photos were arranged in a narrative sequence, beginning with one captioned 'Plumed Egret brooding', running through 'A victim of the plume hunter' and the chicks 'Crying for food', to the final image captioned 'At the last gasp', showing three egret nestlings at the point of death.[8] It was not the first time that photography had been deployed in the service of conservation, but it was among the most emotionally powerful such deployments the world had yet seen.

The RSPB republished the photographs in newspapers, and even employed teams of sandwich-board men to parade the streets

Figure 5. Sandwich-board men hired by the RSPB in London to carry the anti-plume-trade message via Mattingley's distressing photographs.

of London bearing placards emblazoned with enlargements of Mattingley's heart-breaking photos. Beyond that, the RSPB mobilised its international contacts, distributing its pamphlet and photographs into continental Europe and North America. Everywhere, it was a highly emotive campaign, so Mattingley's photographs were eagerly taken up by the crusaders against death-dealing millinery.

In America, protests against the plume trade were led by women, especially the upper-middle-class women of Boston and Philadelphia. Two such women, Harriet Hemenway and her cousin Minna Hall, founded the Massachusetts Audubon Society in 1896 to advance the cause. Audubon Societies rapidly proliferated. They had male members, too, including some in high office, but the drive and energy of the Audubon movement came from women, especially in campaigning against the plume trade. These same women were instrumental in establishing birdwatching as a pastime in late-19th- and early-20th-century America, differentiating their hobby from male-dominated ornithology with its shotguns and arsenic.[9]

In the United Kingdom, the Royal Society for the Protection of Birds, founded in 1889 though without the royal prefix until 1904, was a similarly female-dominated body that was created to fight the plume trade. As in America, these were the years of first-wave feminism, and the leading opponents of the plume trade tended to be drawn from the same social strata as the leading proponents of women's rights and suffrage. Some historians have drawn connections between the two, doubtless with some validity.[10] However, the situation in Australia was somewhat different. There was a strong feminist movement in the late-19th and early-20th centuries,[11] and some Australian feminists were also involved in anti-plume-trade campaigning, but the campaign never became female dominated in Australia as it did in America and Britain.

That said, Australian women certainly agitated against the plume trade. In 1894, a South Australian branch of the Society for the Protection of Birds was founded, primarily to stop the plume-hunters. Like its British parent, its members were predominantly women.[12] Poet Mary Gilmore took a forthright stance against the trade. So did fellow writer Katherine Susannah Pritchard, who, in a 1908 newspaper article, excoriated her plume-wearing sisters:

> The tragedy of the egret is revived by every woman who adorns herself with plumes that have been torn from the body of that shy bird-mother in nesting time.
>
> When women realise their bond with this ill-fated mother, a wave of incensed sympathy will make them discard the light and dainty feathers which they have loved to wear, and make them no longer placid consenters to butchery. The woman who wears an osprey will become a mark for the finger of surprise, indignation and contempt. Her vanity will be proclaimed as something flagrant and inexorable, indifferent alike to appeals for pity and the cries of pain, since she requires the cruellest suffering and rapine to administer it.[13]

As Pritchard's words attest, female campaigners asserted the same ethical injunctions, in the same tone, as their male counterparts. But it was their male counterparts who in Australia took the limelight in protesting against the plume trade.

Perhaps it was for this reason that the Australian anti-plume-trade campaign did not become entangled in opposition to specimen-collecting, as it did in Britain and America. Mattingley himself was a collector, especially of eggs, but also of skins. At the Riverina heronry in 1906, he took more than photographs; he took a clutch of Plumed Egret eggs that he exhibited to a meeting of the Field Naturalists' Club of Victoria.[14] Campaigning against the plume trade in Australia was led by the (R)AOU, with the South Australian Ornithological Association also active in that state. Membership of both organisations was predominantly (though not exclusively) male, and most members engaged in collecting of one kind or another. Though ardently opposed to killing birds for millinery ornament, at least some were concerned that laws to that end should not disrupt their own collecting adventures.

Lobbying for legislative intervention exposed other facets of the campaign against the plume trade. When campaigners pleaded with women to desist from decorating themselves with feathers, their appeals were typically ethical and emotional. When campaigners tried to persuade legislators to act against the plume trade, they more likely advanced economic and utilitarian arguments. So, for example, when a deputation from the AOU met with the federal minister for customs in 1910, to push for legislation outlawing the sale, exchange, import, and export of the plumes and skins 'of certain wild birds', they highlighted 'the value of birds to the community'. Mattingley was a member of the deputation, and he explained to the minister how various birds, including herons and egrets, helped control liver flukes by predating on snails, the flukes' intermediate host. Such birds, he averred, 'were valuable servants of man'.[15]

The 1910 deputation also informed the minister of overseas legislation that Australia should emulate. From its foundation, the AOU had kept abreast of international developments in conservation, looking especially to the US for models of the legislative measures needed to protect birds. Volume one of the *Emu* carried an article on 'Bird Protection in America', informing AOU members of innovations there, and praising American efforts 'towards suppressing the trade in birds for millinery purposes'.[16] Many more such articles appeared over subsequent years, although one by Thomas Stephens in the July 1914 issue of the *Emu* deserves special attention. Stephens lauded two measures taken by Congress in 1913: one protected migratory birds, while the other 'freed the United States for ever from the shame and the horrors of the millinery trade in wild birds' plumage'. He acknowledged that Australian governments had 'done something toward preserving bird life', but maintained that 'the more drastic action of the United States' needed to be duplicated to stop the 'merciless destruction of beautiful, free, wild birds'.[17]

The American legislation of 1913 sounded the death knell of the plume trade. It didn't completely stop the killing of birds for their feathers, but, by shutting down the enormous American plume market, it removed a major incentive for the mass slaughter of birds for hat decoration. Other countries soon followed the American legislative example, further cramping the plume-hunters' business model.[18] Yet while it was legislation that ultimately stymied the plumage industry, that didn't diminish the importance of the public campaigns to change hearts and minds. As Stephens noted in his 1914 *Emu* article, the two crucial measures for bird protection 'were swept through Congress on an irresistible tidal wave of insistent public sentiment'.[19] That tidal wave not only swept through Australia, too, but an Australian birder, Arthur Mattingley, added immensely to its energy.

While early-20th-century bird lovers lobbied hard for protective legislation, they knew that laws alone were insufficient. As Walter Froggatt declared in the first paragraph of his book on *Some Useful Australian Birds*, 'Unless the people themselves are awakened to the beauty and value of our fauna, no Act, however perfect, can be of much use.'[20] Consequently, birders devoted a great deal of effort to educating the public on the worth of birds and to encouraging an appreciation of their charms. Some of that publicity was directed at adults; some at children, most notably in the case of the Gould League of Bird Lovers. For both, the objective was to build bonds between people and birds, thereby dissuading the former from killing the latter.

In the early decades of the 20th century, conservationist campaigning was seldom grounded in ecological science. Yet birders were not blind to the importance of habitats. The South Australian adventurer and ornithological collector Captain S.A. White protested in 1910 against the destruction of 'fine swamps' to make way for farms. Where, he asked, 'will our poor birds go' when 'these swamps, the home and breeding-place of thousands of water-fowl for generations, will be dried up one after another'?[21] However, the retention of habitats was only a minor strand in conservation campaigning in the first few decades of the 20th century. Even those who asserted the importance of habitat-preservation acknowledged that action on that front languished for lack of supporters in positions of power.[22] Bird lovers sensibly focussed their energies on objectives with better chances of success: winning hearts and minds, on the one hand, and curtailing activities directly detrimental to birds, such as shooting and trapping, on the other.

In the 1930s, ecological understandings became more prevalent in pleas for the conservation of Australia's avifauna. At the beginning of that decade, Spencer Roberts, a doctor based in Stanthorpe, Queensland, published an innovative proposal for replacing current arrangements for 'bird protection' with more rigorous strategies of 'bird preservation'. Influenced by the then-novel science of population ecology, he insisted on the need for solid scientific data on the abundance, distribution, and fluctuations of avian populations. These data, he maintained, would elevate bird conservation to 'a different plane'. Roberts' proposed strategies entailed two major innovations: systematising the establishment of sanctuaries to retain the widest possible diversity of habitats, and categorising threatened species into four 'classes' according to the degree of risk they faced and the level of protection they required.[23] Several decades would pass before schemes of this kind were implemented in Australia.

Yet the efforts of old-school bird protectionists were not futile. They did end the plume trade, and the egret population bounced back. In 1927, birder and lawyer Norman Favaloro could report on one of the former main areas of operation of the plume hunters:

> All members of the Union will be pleased to learn of the marked increase in the numbers of Egrets, Herons, and Spoonbills on the irrigation settlements neighbouring the Murray and its tributaries. For some time it was feared that the commercial value of plumes would result in the extermination of our beautiful Egrets, but since ... the plume trade received its death-knock, the birds have gradually increased, and now it is no unusual sight to see Egrets and Spoonbills on almost every Channel bank.[24]

Mattingley's distressing photographs, which were taken close to the place on which Favaloro reported, eventually had an impact beyond anything he could have imagined when he hauled himself and his camera out of a swamp and up a tree to an egret's nest in 1906.

CHAPTER SIX

Crested Shrike-tit: Amity

THE CRESTED SHRIKE-TIT IS a striking bird: bright yellow below, olive above, with a boldly patterned black-and-white neck and head, topped off with a short crest that gives the bird its name. Small but robust, it has an energetic, business-like manner as it forages through the foliage. Though colourful, shrike-tits can be unobtrusive, especially when high in the trees, which is where they usually are. Their presence is often given away by their call, a soft and mournful whistle, or by the sound of ripping bark as they tear shreds off the tree-trunks with their strong, hooked beaks in search of insects and spiders. Typically, they nest high in the tree-tops, but occasionally choose a site lower down.

In 1914, Alec Chisholm found such a lower-sited nest near his hometown of Maryborough in Victoria. Seizing the opportunity for a photograph, he began preparations. Although the nest was low for a shrike-tit's, it was still about five metres aloft, and, lacking telephoto lenses, the camera and its operator had to be elevated to that level. So Chisholm and his birding companion improvised an

Figure 6. Chisholm's cumbersome photographic contraption.

enormous tripod out of borrowed bits of timber, mounted a camera on its summit, and set up a stepladder on a horse-drawn cart inside it, enabling Chisholm to clamber to his precarious photographic perch. But the discomforts and dangers of this contrivance were just the beginnings of his tribulations in photographing the shrike-tit.

He made three attempts to photograph the birds at nest. The first had to be abandoned, because he and his birding companion miscalculated the required length of the tripod's legs and had to rebuild it from scratch. Next day, they tried again, but this time the wind was blowing strongly and the nest swaying too much for his camera to cope with. For the third attempt, the weather was perfect, but on climbing to his photographic post, Chisholm discovered that a predator had robbed the nest. Or, as he put it in his characteristically lavish style, 'Some winged Assyrian had been down on the fold over-night and despoiled the nest, thereby bringing as much chagrin to a pair of spry youths as sorrow to a pair of brave birds!' Recapping his unsuccessful efforts, he recounted that, 'Earlier in the hunt for Shrike-Tits' nests I had endured with equanimity a fall from a tree-top, the breakage of a camera, and sundry other "incidentals", but this latest failure was the greatest trial of them all'.[1] Instead of photographing the birds and nest, he persuaded his younger brother, Norman, to take a photo of his cumbersome photographic contraption.

That photograph became the frontispiece for Chisholm's first book, *Mateship with Birds*, published in 1922. It's a wonderfully apt title. In it, Chisholm urged Australians to open their hearts to their avian compatriots: to bond with the birds around them, loving and cherishing them as fellow members of the national community. Overtly and intentionally emotive, the book was written in a lush, romantic style, suffused with anthropomorphism, and studded with stanzas of quoted verse. With characteristic passion, he entreated Australians to accept 'the wing of friendship' the birds were holding out to them, for if they did, 'the story of St. Francis of Assissi [sic] would pale by comparison with our experiences of every day'.[2]

The poet C.J. Dennis—Laureate of the Larrikin, as he was dubbed—wrote the introduction to *Mateship with Birds*. Lauding Chisholm's 'fraternal attitude' toward birds, Dennis explained that, 'Many a learned savant shoots birds with a gun and writes about them as a pedant. Mr Chisholm shoots them with a camera and writes about them as a human being.'[3] The book struck a chord with the public, drawing appreciative letters from readers, and accolades from reviewers. One reviewer commended the author for writing 'as an authority on birds, but not as one who sacrificed his soul to the scientific aspect of his hobby. Chisholm is essentially a student-lover of Nature, and for such as he "the poetry of earth is never dead".[4]

At almost the same time as *Mateship with Birds*, two other delightful bird books were published: *Birds of our Bush, or Photography for Nature-Lovers* by R.T. Littlejohns and S.A. Lawrence (1920), and Les Chandler's *Bush Charms* (1922). All three books bear witness to the transformation that birdwatching was undergoing around that time, especially the eschewal of skin- and egg-collecting in favour of field identification and observation. Aligned with and reinforcing that transition was the growth of bird photography and a strengthening demand that studying birds cause them no harm. All three books were, to varying degrees, romantic in style, forthright in their emotionality, and unapologetic in their aesthetic enthusiasm, while at the same time they meticulously documented the lives and behaviour of birds.

Was it coincidence that these books were published just after World War I? Or was their rejection of shooting birds connected with revulsion against the killing fields of war? The only one of the four authors whom I know to have served in the war was Les Chandler, who enlisted in the Army Medical Corps in July 1915, aged 27, and spent most of the next three years on the Western Front. He was gassed in April 1918. On almost the same day, his army mate and fellow nature-lover Morry Thompson was killed in action. The two men had planned to write a book on birds when they returned to Australia.[5] *Bush Charms* was a residue of that plan.

The horrors of war occasionally surface in *Bush Charms*. In a passage condemning the slaughter of Spotted Bowerbirds by orchardists as punishment for the birds' alleged fruit-stealing habits, Chandler lamented:

It is a sad sight to see the old neglected bowers falling into decay, and the crude playthings of these marvellous birds crumbling into dust. Visions of the smashed villages of France and Belgium where the toys of little children lay broken and scattered in the wreckage, rose before me. No longer would happy birds fill in the care-free hours at play in these bowers, and my heart was saddened to think that man could not place the aesthetic tastes and clever performances of these birds above the greed of gain.

In other passages, he explained how he 'longed for the peace and charm of our Australian bush' as an antidote to the carnage he witnessed in Europe.[6]

Littlejohns and Chisholm did not enlist; I am uncertain about Lawrence, as biographical information on him is scant, but the little available evidence suggests he did not. However, the war impacted everyone, not just those who experienced the bloodshed and devastation at first-hand. For Australia, as for all combatant nations, World War I was a culturally traumatic event.[7] Killing birds for ornithological study had been contested since the turn of the 20th century, and it is plausible to suggest that the bloodletting of war fuelled revulsion against the bloodshed of collecting.

All four authors, in their youth, had shot birds and collected their eggs, but had forsaken those practices long before they published their first books. Chandler represented this more as a personal choice than as a moral imperative for all to follow:

I am glad to say that I have abandoned for many years the collecting of specimens. The art of the camera, and the charming trustfulness of many of the feathered creatures, won me over to a different view of collecting, and created a sentiment for all wild creatures that was more in keeping with the aims of a true nature lover.[8]

His book's introduction by Charles Barrett, one of Australia's earliest opponents of collecting, put things more forcefully. Praising Chandler for having 'laid aside his gun, and ceased to be a collector', Barrett added that, 'Now, he is known as a leader among naturalists who work with field-glasses and camera, and reap a richer harvest than those who "study" bird life with a gun. The new way is the better way.'[9]

Of the three books, Littlejohns and Lawrence's was the most forthright in condemning collecting. It was also the most explicit in proffering photography as an alternative:

> Our chief argument in favour of photography as a means of observation ... is that the photographer, in gaining his ends, need leave no trail of destruction in his wake. The collector, whether he shoots the birds or takes their eggs, has only desolation in one form or other upon which to pride himself should he contemplate the result of his day's work. The photographer, on the other hand, leaves, or should leave, his subjects just as he found them, and no worse for his interference. He may go back again if he wishes, to observe the progress of his friends or to picture some new phase of their lives.[10]

Some readers today may find the wording banal, but in 1920 it was a manifesto for a new era.

Littlejohns and Lawrence began their book by declaring that they had 'little knowledge of what is generally accepted as scientific ornithology', because 'the problems of nomenclature and sub-division of species interest us not at all'. In 1920, those 'problems' were at the forefront of ornithological science, which in Australia had been plunged into an exceptionally disputatious phase under the influence of Gregory Mathews, who had an insatiable appetite for splitting species and discovering new subspecies. This was the kind of bird study that Littlejohns and Lawrence lampooned:

> The variations ... which distinguish these subspecies are so slight in many instances that a most minute examination is necessary to determine the identity of the specimen—hence its untimely end. Just where the advantage of all this investigation comes in we, as laymen, are unable to say. Certainly we can point to a serious disadvantage in the amount of slaughter entailed.[11]

The young photographers seem to have been thumbing their noses at the ornithological establishment, rather in the manner that a later generation of recreational birders, calling themselves 'twitchers', scoffed at their staid and starchy birdwatching predecessors.

Birds of our Bush began with a chapter aptly titled 'On Collecting with a Camera', flagging both a break from traditional collecting with a gun and the continuities between new-fangled photography and old-fashioned shooting. The photographer collected images, not carcasses, but still sought keepsakes of the birding encounter, photographs replacing the skins and eggs of the collector. Like skins and eggs, too, photos were valued for their educative potential—Littlejohns and Lawrence acclaiming them 'a most efficient and easily distributed means of education' about birds. They also treasured photos for the memories they preserved and the associations they evoked.[12]

Chandler, too, represented photographs as *aide memoires* for past interactions with birds, although he put this in the context of the camera's inability to fully capture the wondrousness of nature. Celebrating 'the companionship of wild birds', he added that:

> The eyes see much that cannot be reproduced on a photographic plate. Color and atmosphere are missing in the black and white print. It is like plucking a wild-flower and placing it in a vase. One has the flower, but the major portion of the picture that so delighted the eye is left in the bush. So it is with a photograph; the bird and nest are shown, but perhaps a leaf in the background, burnished to exceptional beauty by a ray of sunlight, shows as a white, out-of-focus blob. Still, the picture may depict a pretty phase of the bird's homelife, and, in later years, it helps to reconstruct in one's mind a series of interesting incidents that attended the taking of the photograph.[13]

Littlejohns and Lawrence were less equivocal in singing the camera's praises. Their book, after all, was an exercise in proselytising for bird photography as a popular pastime, whereas Chandler strove to foster love of nature on a broader front.

As part of their promotion of bird photography, Littlejohns and Lawrence persistently reiterated the fact that their photographs had been taken with cheap cameras and that the hobby was inexpensive and accessible to all. They did not shy away from the challenges of photographing birds, instead presenting them as adventures, sometimes in a rather boyish fashion. Thus, for example, they illustrated 'photographing in an elevated position' with a photo of themselves on a makeshift structure assembled from bush timbers and looking even more rickety than the contraption that Chisholm had used to try to photograph the Crested Shrike-tits' nest. Throughout the interwar years, photographing birds continued to be as dare-devil and dangerous as collecting their eggs. Bird photography was not for the faint-hearted.

Chandler was among the first in Australia to use hides for bird photography. Littlejohns and Lawrence acknowledged the usefulness of hides, but did not use them. They were, nonetheless, 'advocates of bird-photography from very close range', recommending 'eighteen to twenty inches as a satisfactory distance between the lens and a small bird'. Before the advent of telephoto lenses, getting physically close was the only way to secure a close-up image. Littlejohns and Lawrence suggested that this closeness helped the photographer, and ultimately the viewer of the photograph, to connect with nature, maintaining that:

> [M]uch of the charm of Nature Study arises out of the fact that the observer sees and knows the wild things of the bush more closely than the average person, and that from this very nearness there becomes apparent a great deal of beauty which is lost from a greater distance. Our contention is that a photograph, if it is to fully realise its possibilities, must reproduce this sense of nearness.

Seeking to convey a sense of intimacy with nature, they insisted that photographs must be 'genuine representations of the wild bird'.[14]

Yet Littlejohns and Lawrence engaged in practices that seem at odds with this injunction, at least by today's standards. They tethered young birds as decoys to attract the parents, and blocked the entrances of pardalotes' nesting tunnels to force the birds into photogenic poses. Immediately after declaring as their motto 'Never descend to taking pictures of subjects which are not entirely natural', they inserted a photograph of a nesting Eastern Yellow Robin being stroked by a human finger. More striking still is their photo of a female Mistletoebird perched on an index finger, feeding a fledgling perched on the middle finger of the same hand. They acknowledged that they had removed the youngster from the nest, but nonetheless offered 'the assurance that every bird pictured in these pages was a wild one, and was taken in natural surroundings'.[15]

Similarly, Chandler expressed his delight in having a Yellow-faced Honeyeater feeding her chick while both birds perched on his fingers, and his joy in an Eastern Yellow Robin allowing itself to be stroked while sitting on its nest.[16] He, too, deliberately arranged birds at nest. Reporting on a photograph of a young Crested Shrike-tit that he published in *Bush Charms*, he admitted that, 'The young bird in the illustration had left the nest, but I caught him and placed him on the rim to show the beauty of nest and bird in one picture. It is not a natural picture, and I only place it in this book on account of the rarity of the subject.'[17] The final sentence suggests that Chandler may have had a more rigorous conception of what constituted 'natural' in the new world of bird photography than did Littlejohns and Lawrence. Nonetheless, the nonchalance with which Chandler recounted such stratagems indicates that he saw nothing wrong with them.

Nor did Chisholm. Although he failed in his 1914 attempt to photograph Crested Shrike-tits, at some time between then and 1922 he had success in the endeavour, so was able to include in *Mateship with Birds* a photograph of a shrike-tit's nest with two eggs, and another of a male shrike-tit perched alongside one of its chicks. Like many bird photos at the time, the latter was staged. Chisholm explained that he chased the young shrike-tit 'until the poor wee thing came to the ground in sheer weariness':

Then it was placed upon a low, horizontal bough, and tied down lightly with a piece of cloth (torn from the lining of my coat).

And there the crested infant sat, "Charr-charring" wonderingly at the great world in general and its strange captor in particular, until its father flew down and sat alongside. That is how one of the accompanying photographs came to be taken.[18]

Other photos in the book show people touching birds, children standing within inches of occupied nests, and a fluffy baby Logrunner on a human hand.

Some birders today may fling up their hands in horror. But past practices must be understood in historical context. Chandler, Chisholm, Littlejohns, and Lawrence were trying to advance new ways of birding that entailed no loss of birdlife. Photography was part of the new mode that they hoped would displace the gun. Not only did the camera not kill the bird; it produced images that could encourage people to appreciate birds and value them as living beings. A photograph was not only a memento that carried a freight of memories for the photographer; it was also a vehicle conveying the glories of birdlife to the wider public, thereby promoting the conservationist cause. As these young photographers realised, the message was best conveyed by close-up images and depictions of intimate encounters between birds and people.

In these books, stroking birds and having them perch on fingers were lauded as expressions of trust. Chisholm captioned his photo of a Pale-yellow Robin about to be touched by a human finger as 'a study in trustfulness'. His photo of two boys gazing at a nesting robin from about half a metre away he captioned as 'Mateship with Mother Robin'. Intrusion into the bird's domain was accepted as part of the process of building trust or fostering mateship. After all, such intrusion was intrinsic to bird photography: it was simply impossible with the available cameras to get a good, clear photo of a bird without getting up-close and personal with it. Using a hide, as Chandler did, may have moderated the intrusiveness, but did not eliminate it. Photos that captured a bird's personality were especially prized for their capacity to emotionally draw the viewer into a positive relationship with birds. Such photos could be taken only by invading a bird's personal space, usually when it was engaged in the demanding and delicate duties of nesting.

The fundamental message flowing through all three books was that humans and birds should be bonded in amity. We should cherish them; we might observe them, study them, photograph them, but we must not wantonly kill or injure them. Their authors endorsed—even encouraged—a level of interference that has since fallen into disfavour, but their aim was to bring people closer to birds, emotionally and aesthetically.

Birders today might decry what Chandler and Chisholm did to the Crested Shrike-tits: capturing them, tethering them, poking cameras into their nests, even erecting huge photographic contraptions that advertised the nest's whereabouts to every passing predator. But these were important steps in the decoupling of watching and photographing birds from killing and collecting them. As Littlejohns and Lawrence's chapter 'On Collecting with a Camera' attests, what was happening was not so much the invention of a new hobby as the reconfiguration of an old one, the shedding of the more brutal elements allowing more amicable aspects to come further to the fore. Ultimately, the change was massive, but, like most successful transformations, it proceeded piecemeal.

CHAPTER SEVEN

Laughing Kookaburra: Identifying

THERE CAN SCARCELY BE a more iconic Australian bird than the kookaburra. Its image has advertised a dazzling array of products, from Metter's Early Kooka stoves to Kookaburra Brand hosiery. From 1929 to 1975, Movietone newsreels at Australian cinemas began with a pair of kookaburras laughing boisterously at nothing. In their exposition of Australian bird names, Ian Fraser and Jeannie Gray remark that, 'This species has possibly attracted more folk names, many of them affectionate human names, than any other Australian bird.' Among them were 'Jacky', 'Jacko', 'Laughing John', 'Breakfast Bird', and 'Settler's Clock'.[1]

It's the Laughing Kookaburra that's familiar to most Australians. Across the north, there's another species, the Blue-winged Kookaburra, with a brighter blue wing-panel, a staring white eye that can give the bird a crazed appearance, and a matching call that sounds more like a maniacal cackle than a laugh. Reputedly, some early British colonists found the call of the laughing species repellent or demonic, but it quickly became assimilated into the colonial psyche and appreciated as a characteristic sound of the bush. The kookaburra's profile was boosted by its reputation as a snake-killer—a reputation far exceeding the bird's accomplishments in that regard. In any event, Laughing Kookaburras came to be regarded with such affection that they were deliberately introduced into Tasmania and south-western Western Australia, where now they thrive.

The kookaburra's only rival as an Australian avian icon is the Emu, featured on the national coat of arms and in other emblematic sites. But Emus, as wild birds, are unfamiliar to most Australians today, having retreated from more populous parts of the continent into the dry interior. Kookaburras, by contrast, have adapted so well to the changes wrought by colonisation that today they laugh as raucously

Figure 7. Pioneers of the Australian field guide: John Leach (front); Neville W. Cayley (behind).

and as plentifully in suburban gardens as in the bush. They're among the wild birds most frequently fed by urban Australians and among the most familiar. Few Australians, even those with no interest in birding, could not instantly identify a kookaburra. I suspect that's true even in the Northern Territory, which lacks the laughing species, but does have the blue-winged kind.

Presumably, it was for the bird's iconic status—certainly not for any intricacies of identification—that Neville Cayley chose to adorn the cover of his 1931 field guide, *What Bird Is That?*, with a Laughing Kookaburra perched on a question mark. It stamped the book as indubitably Australian, which its title did not; Frank Chapman had published an American guide under the same title in 1920. Cayley's was Australia's second field guide, having been preceded 20 years earlier by John Leach's *An Australian Bird Book*, whose novelty was announced in its subtitle: *A Pocket Book for Field Use*.

The advent of the field guide, in Australia as throughout the Western world, helped drive birding practice away from the gun and collector's cabinet, toward binoculars and field observation. Field guides encouraged people to take pleasure from seeing and identifying birds in the wild: from learning, as Leach put it in a nicely personal turn of phrase, how 'to name the birds they meet'.[2] Beyond that, the very format of field guides—essentially illustrated catalogues of birds—invited people to tick off those they'd seen, thereby promoting the practice of listing that has since ballooned into a popular birders' sport. Although Leach presented birdwatching as a species of edifying recreation infused with science and self-improvement, the field guide genre, which he introduced into Australia in 1911, also pushed the pastime toward simply having fun.

Before field guides, there were handbooks and keys for identifying specimens, but, being meant for use on stiff bundles of feathers and skin held in the hand, they gave little help in ascertaining the identity of flighty creatures glimpsed in the treetops. Unease at the morbid side of such works was voiced even by their authors. In his 1905 'Dichotomous Key to the Birds of Australia', A.G. Campbell, son of A.J. Campbell, appealed to fellow ornithologists to 'give any advice that may improve the Key, especially with the view of making it *applicable to work in the open*, without the destruction of bird life'.[3] Campbell evidently liked the idea of a field guide, and hoped his own Dichotomous Key might advance the development of one. Growing sensitivities over the slaughter of birds for ornithological study were pushing birders toward devising means of identification that did not involve killing the subjects of their scrutiny.

Killing birds to identify them was so common that Leach felt obliged to explicitly censure it. In his *Descriptive List of the Birds Native to Victoria*, a 1908 precursor to his *Bird Book*, intended for use in schools, he stated:

> It is earnestly hoped that no bird will be killed for the purpose of identification, but that teachers will note the approximate size, class of country the bird lives in, any particular markings. on it, and any peculiar habits it possesses. Then turn to these notes .... In most cases, in the field, a ready identification can be made, and frequent observation will enable one to pick out the white eyebrow, the tiny red spot, or the faintly-streaked chest, &c., which enables the skilled observer to be sure of the identity of our feathered friends.[4]

Leach's words show his awareness of what later became known as field marks, but in his *Bird Book* he failed to draw attention to them as identification aids. Explicit attentiveness to field marks did not appear in Australian field guides until the 1940s.

By the time Cayley published *What Bird Is That?*, the transformation of birding into a bloodless outdoor recreation was well underway. In the introduction, Cayley announced that the 'idea is gradually being discarded, that, to study birds one had to make a collection of skins and eggs. Nowadays the camera has displaced the gun, and the photographic album, the skin and eggs cabinet'.[5] He was right about the trend, although perhaps a little optimistic about how far it had progressed by 1931. He might also have added the field guide to the devices that were displacing the gun from the birder's toolkit. His own, as well as Leach's, were potent contributors to that displacement.

Leach's and Cayley's field guides were destined to become Australian classics, but they were very different books. Despite the promise of its title, Leach's book did not cover the entire continent's avifauna, but was restricted to those found in Victoria. So, while it described and depicted the Laughing Kookaburra (under the name

Figure 8. Monochrome illustration from J.A. Leach's *An Australian Bird Book*, 1911.

Laughing Kingfisher), it omitted the blue-winged species. Not until the fifth edition of 1923 were non-Victorian species added to the book, and then merely by tacking a 31-page supplement onto the end, listing 301 species from outside Victoria with minimal descriptions, few illustrations, and a separate index. Here, the Blue-winged Kookaburra made its first appearance, with no illustration and a description merely stating that it was 'like' the Laughing species, but with 'blue patch on wing; tail blue; f. tail red barred blue'. In subsequent editions, through to the ninth published in 1958, Leach's guide never quite managed to transcend its Victorian origins.

Cayley's *What Bird?*, by contrast, had continent-wide coverage from the outset. Indeed, it was the first work to depict all the known birds of Australia in a single volume, and all the depictions were in full colour. Cayley was an accomplished artist, and his illustrations were vastly superior to those in Leach's guide. Even so, they were not adequate to the immense demands that Cayley placed upon them, for he provided no verbal descriptions of plumage at all, relying entirely on the plates to convey a species' appearance (except for introduced species, which were segregated into an appendix and described, but not illustrated). Exacerbating the shortcomings of this reliance on illustrations, the plates were crowded, each bird's picture varying between small and tiny.

While Leach's *Australian Bird Book* was avowedly a field guide, its contents ranged far more widely than is now usual for that genre. Over half its text was devoted to 'A Lecture' that traversed a vast array of topics, most of which had no relevance to identification. Indicative of birding's origins in the popular natural history of the Victorian era, the lecture's miscellany of topics included speculations on the configuration of Earth's continents, discussion of the ideas of Alfred Russel Wallace and Charles Darwin, assessments of the size of the mutton-bird harvest on Cape Barren Island, musings on the purpose of the double-storeyed nest of the Yellow-rumped Thornbill, comparison of the musical talents of British and Australian songbirds, and reflections on the evolutionary significance of coloured wing-patches. American field guides at this time, when the genre was in its infancy, also typically carried a substantial freight of general natural history.[6] Leach's guide exceeded them by several orders of magnitude.

Bird species in Leach's book were listed in taxonomic order, a practice that has become a field guide convention. However, he provided details on avian systematics, in both lecture and species notes, that went far beyond anything useful for identification in the field. Taxonomy and systematics were then at the forefront of ornithological science, and were Leach's own specialist interests. A year after his *Bird Book* was published, he was awarded a Doctor of Science by the University of Melbourne for a thesis on those topics. Yet the author of Australia's first field guide had limited expertise in the field. 'Scarcely a field ornithologist' was how Leach was characterised by Tom Iredale, himself a leading systematist. Fellow birder Charles Bryant described Leach as 'largely a cabinet man, concerned with systematic and taxonomic ornithology'.[7]

Although exceptionally scientifically credentialed by the standards of the day, Leach was not a professional ornithologist. He was a schoolteacher who by 1911 had risen to the rank of inspector in the Victorian Department of Education. Yet while not a professional ornithologist, the term 'amateur' scarcely fits him either, for his employment was primarily in the domain of nature studies, with evident connections to both his academic qualifications and his ornithological interests. He can be seen as straddling an amateur–professional divide that was neither deep nor unbridgeable. Such straddling was then common. It remains so in bird studies even in today's world of rampant credentialism.

Cayley lacked Leach's scientific qualifications, and came to ornithology via art. It ran in the family: his father, Neville Henry Cayley, was also a bird artist, although he specialised in painting game birds for a clientele of sporting hunters.[8] The shift of emphasis from father to son—from celebrating the hunt to enhancing field observation—is emblematic of a bigger shift in the Australian people's relationship with birds. *What Bird Is That?* would push it along further.

Unlike Leach's, Cayley's guide stripped the natural history content to a minimum, and the taxonomic content to even less. Against the usual field guide practice of listing species in taxonomic order, Cayley arranged his primarily by habitat and secondarily by size. This had twin drawbacks: similar species were often separated by many pages, and birds in the wild commonly strayed from the habitats to which Cayley allocated them. Exacerbating that problem, his nominated habitats were sometimes misleading. For example, the White-browed Crake is listed and illustrated under 'Some Birds of the Mangroves', whereas an observer is far more likely to find the species in wetlands with waterlilies.

Although Cayley included far less natural history content, on one topic he gave far more information than Leach. Cayley described nests and eggs in detail, usually devoting between a quarter and a third of the words on each species to these topics. Leach's species descriptions, by contrast, said nothing at all about nests and eggs, avoiding those topics 'in case schoolboys and scouts should be tempted to lay up treasures of eggs', as a wry reviewer in the *Emu* put it.[9] Two years before publishing his *Bird Book*, Leach had been a founder of the Gould League of Bird Lovers, one of whose aims was to quash the then-common children's hobby of egg-collecting. Since his book was aimed partly at children, he deemed it inappropriate to include anything that might seem to sanction the practice.

Cayley lacked such qualms, although he, too, was an active member of the Gould League, and in the 1930s egg-collecting was still a popular schoolboy pastime. Moreover, publication of *What Bird Is That?* was sponsored by the Gould League of Bird Lovers of New South Wales, and, from 1935, the lucrative royalties from its sale went to the league. Presumably, Cayley included material on nests and eggs to cater to popular interest. Although egg-collecting was declining as a hobby, at least among adults, people still delighted in the intricate architecture of nests and the lustrous geometry of eggs, and sought to admire those marvels in the wild. Besides, nests were where birds could be photographed. Over the 20 years between Leach's and Cayley's guides, the popularity of bird photography had grown immensely, and information on birds' nests was helpful for practitioners of that pastime.

Although he sidestepped the turmoils of taxonomy, Cayley provided translations of the scientific name of every Australian bird species. So, for example, he explained of the Laughing Kookaburra, *Dacelo gigas*, that its generic name was an anagram of *Alcedo* (Latin for kingfisher) while *gigas* derived from the Greek word for 'giant'. It seems a perfectly good name; 'Giant Kingfisher' was once among the species' popular names. The bird had the same scientific name, although a different vernacular name, in Leach's book. But as is the way with scientific names, *Dacelo gigas* didn't endure. The Laughing Kookaburra is now *Dacelo novaeguineae*. It is found nowhere in New Guinea, but the rule of priority reigns supreme in scientific nomenclature, and since *novaeguineae* was the first specific name given to the bird (by Johann Hermann in 1783, in the mistaken belief that the specimen had been collected in New Guinea) that is the name fixed to the bird today.

The scientific name of the Blue-winged Kookaburra, by comparison, has been a model of sense and stability. In Cayley's guide it was *Dacelo leachi*, as it was when it finally made it into Leach's book in its fifth edition, although the Leach after whom it was named was not the author of this guide, but the early-19th-century zoologist William Leach, who had concocted the anagrammatic generic name *Dacelo*. In today's field guides, the Blue-winged Kookaburra is *Dacelo leachii*, varying from earlier nomenclature only by an additional i.

Both field guides were publishing successes. By the time of Leach's death in October 1929, seven editions of his *Bird Book* had been published, with sales totalling over 30,000 — 'surely a record for nature-study books', according to fellow naturalist Dr Brooke Nicholls.[10] Even Leach's death did not staunch the flow of new editions: an eighth edition came out under Charles Barrett's editorship in 1939, and a ninth in 1958 with Crosbie Morrison as editor. All editions except the ninth retained the full text of Leach's lecture with only slight amendments, and the ninth merely varied this by providing a revised version of the lecture under the title 'Australian Bird Life'. A hefty chunk of natural history seems to have been considered an essential component of what was still presented as 'a guide to identification' that would help people 'name the birds they meet'.

What Bird Is That? had an even more illustrious career. By Cayley's death in 1950, it had sold 50,000 copies and had gone through 13 reprints, with several more to follow before the second edition was issued in 1958. This was edited by three of Cayley's friends, all renowned birders: Keith Hindwood, Alec Chisholm, and Arnold McGill. According to McGill, Hindwood had been a major contributor from the outset; indeed, he claimed that 'almost all the text [of the first edition] was Keith's work'.[11] The second edition made some improvements, for example in the quality of reproduction of the colour plates and in giving more information on calls and songs. However, the continuities are more evident than the changes.

Species were still arranged by habitat; no verbal descriptions were given; and no attempt was made, in illustrations or text, to highlight the distinctive field marks that help differentiate similar species. The last is particularly noteworthy since McGill himself had explicitly drawn attention to the value of 'field characters' in a 1952 *Field Guide to the Waders* that he co-authored with Herb Condon. Somehow, this important innovation failed to influence his revision of Cayley's classic a few years later.

Despite its shortcomings, *What Bird Is That?* retained a place in birders' hearts even after Australian field guides underwent a thorough shake-up from the 1970s onward. New editions kept being issued, some making it more user-friendly as a field guide by reducing its size, others going in the opposite direction and inflating it so enormously as to render it virtually impossible to haul into the field. Notable in the latter category is a 1984 deluxe edition revised by Terrence Lindsey, which measured 225 x 310 x 60 millimetres, and weighed three kilograms. It was clearly not intended as a guide for field use, but rather as a commemoration of a field guide for which many birders held fond memories. The text, though revised to some extent, retained the style, format, and most of the contents of the 1931 and 1958 editions. As in those earlier editions, there were no verbal descriptions of plumage, but there were translations of all scientific names and detailed descriptions of nests and eggs—features that give the book a decidedly dated feel, apparently deliberately.

The big innovation of the 1984 edition of *What Bird Is That?* was its inclusion of a whole new suite of Cayley's bird paintings. He had done them for a project that he called, with good reason, his 'big bird book', but which never reached fruition. These paintings, which covered all known Australian bird species, show a more accomplished level of artistry than those in the original *What Bird*

Is That? Their eventual publication in 1984 was a tribute to both Cayley and the field guide he had created, but the fact that they were published in a commemorative edition, totally unsuited to field use, indicates that *What Bird Is That?* had passed beyond the domain of the field guide and into a realm of birding nostalgia. It's the only Australian field guide so far to do so.

According to a publisher's note in a recent reissue of *What Bird Is That?*, 'Cayley's dream was to teach Australians more about their extraordinary birds, and by so doing create a nation of bird lovers to protect them'.[12] That was Leach's dream before him. And it's been the dream of field guide authors ever since. In some guides, such as Leach's, the conservationist message was persistent and pervasive; in others, it was more muted, though even in those it's not difficult to discern, between the lines, an appeal for birdlife to be cherished and protected. After all, authors wrote their guides not only to help birders pin a name to a bird, but also to encourage people to appreciate the birdlife around them. As Keith Hindwood explained in an obituary for his friend, Neville Cayley created his works to express 'his love of birds' and his desire for others to love them, too.[13]

Field guides serve more than instrumental purposes. Because naming is a step toward knowing, and beyond that to appreciating and understanding, field guides open our eyes to the wonders of nature. They offer an entry point into the diversity of life on this planet, and a starting point for making that diversity comprehensible. More than that, they help us bond with nature. Naming the bird is just a first step toward appreciation, but for most of us it's a necessary step. Affection for the kookaburra is so strong, I suggest, because it's a bird almost every Australian can name.

CHAPTER EIGHT

Paradise Parrot: Tragedy

THE PARADISE PARROT MUST have been a gorgeous creature. Zoological collector John Gilbert, who first made it known to science, was so captivated by its beauty that he requested the species be named after himself. His employer, John Gould, declined. Instead, he named it *Psephotus pulcherrimus,* which roughly translates as 'multicoloured and superlatively beautiful'.

Gilbert collected the first specimen on the Darling Downs in 1844. The bird was reasonably numerous there then, and in adjacent parts of inland south-east Queensland. But not for long. The environmental transformations wrought by colonists, probably especially the changed fire-regimes consequent upon Aboriginal dispossession, were deadly to the Paradise Parrot. The wooded grasslands that were home to the parrot were also the lands most coveted by incoming cattle and sheep graziers, whose enterprises imposed new templates of land management.

Trapping for the aviary trade added to the species' plight. Their beauty made Paradise Parrots highly prized cage birds, not only in Australia, but in Britain and continental Europe as well. In all places, they adjusted poorly to captivity, so captive birds commonly made the transition from aviary to parlour, where, stuffed and mounted, they met late-Victorian standards of tasteful decoration. Extraordinarily difficult to breed in captivity, for the parrot the aviary trade was a one-way journey toward oblivion.

By the turn of the 20th century, the species had dwindled to the point that many feared for its survival. With no confirmed sightings of Paradise Parrots in the first two decades of that century, its extinction was widely presumed.

In 1917, Alec Chisholm, then a 27-year-old Brisbane-based journalist, began investigating whether the species was indeed extinct. After four years of false leads and mistaken identifications, he finally received heartening news. On 11 December 1921, a grazier named Cyril Jerrard identified a pair of Paradise Parrots on his property near Gayndah in Queensland's Burnett District. He promptly and proudly told Chisholm of his find.

If such an avian rediscovery occurred today, Gayndah would be instantly flooded with thousands of twitchers, birdwatchers, ornithologists, and onlookers, all bristling with binoculars, telescopes, cameras, and other high-tech paraphernalia. Things moved more sedately 100 years ago. The parrot's rediscovery was not made public until July 1922. Chisholm (who initiated the publicity) took ten months to visit the site, and he was the only birder who did so. His visit lasted only two days, and by the standards of today's birders he was very meagrely equipped. His kit comprised no more than a pair of field glasses and a notebook. Additionally, he held a wealth of skill and experience in finding and identifying birds by both call and appearance: those requisites of birding have endured through all the changes of technique and technology since.[1]

They stood Chisholm in good stead. After a futile first day

spent trudging across the dusty, dry paddocks without a glimpse of a Paradise Parrot, on the second he heard an unfamiliar call and traced it to a tall eucalypt about 150 yards away:

> A sprint down the road followed. What a relief it was, then, to see a female Paradise Parrot perched contentedly aloft! There was no mistaking the slim, graceful form. Presently, she flew into the next tree.
>
> A moment later I gasped with delight, 'Oh, you little beauty!' For there was the male Parrot, the gorgeous red of his underparts gleaming in the midday sun.[2]

Chisholm retold this encounter many times, with slight variations, but always involving a combination of excitement at seeing a rare bird, admiration of its beauty, and remarks on its call and behaviour. It's a combination typical of birdwatchers' commentaries on their observations.

Although an accomplished photographer, Chisholm took no camera to Gayndah. His host, Cyril Jerrard, had not reported the parrots to be nesting at that time, so the unwieldy cameras of the day could only have been an encumbrance. Besides, Jerrard had already photographed the birds.

It took diligence and determination. In March 1922, having found the parrots' nesting tunnel in a termite mound, Jerrard constructed a hide about two metres away, using rough-cut stakes and old hessian bags. There he squatted uncomfortably, waiting for the birds to alight. In somewhat quaint language, he recounted taking the first-ever photographs of Paradise Parrots:

> It was a hot afternoon and my place of confinement was small and ill-ventilated, and in consequence, it was not long before

Figure 9. The hessian hide from which Cyril Jerrard took the first—and only—photographs of Paradise Parrots.

I was "larding the lean ground" (like Falstaff) with moisture from my person. Ere an hour had passed, however, there came a magic sound that banished all sensations of discomfort and made me hastily draw the shutter of my camera and grasp the release, while simultaneously I peered through the interstices of my shelter.

... It was one of the supreme moments of my life. I pressed the release, and at the slight click [the male parrot] hopped back on to the fence. But he was not really alarmed, and I had barely time to change the plate before he was back on the mound. I waited. The female had now come into view on the fence. The male approached the nest hole, just where I wished him to pose, uttered a sweet inviting chirp to his mate and peered into the hole. In answer, as it seemed, to her lord's reassuring word, the female alighted on the summit of the mound. Oh kind Fortune!

I "fired" again, both birds posing for just the instant required. I felt sure I had them clear and sharp, and so it proved when the plate was developed.[3]

I'm unsure how many photos Jerrard managed to take, but over several days he secured fewer than half a dozen. And he was a skilled photographer.

Jerrard photographed the parrots at nest, but their breeding attempt had no happy ending. Keeping the nesting mound under

Figure 10. Male Paradise Parrot at nest-hole in a termite mound; photographed by Cyril Jerrard in March 1922.

Figure 11. Female (top) and male (below) Paradise Parrots: one of only four or five photographs of this species ever taken, all by Cyril Jerrard.

scrutiny, by early April he realised that incubation had gone awry. Once it was clear that the nest had been deserted, he opened the termite mound and found all five eggs had addled. It was a dismal denouement to the hopes raised by his first sighting a few months earlier—and a portent of things to come.

For some years afterwards, Jerrard and his neighbours sporadically saw pairs of Paradise Parrots around Gayndah. Once, a flock of nine was seen feeding in a millet field. But the population was tiny. Jerrard last saw a Paradise Parrot on 13 November 1927, and the last sighting he reported to Chisholm was by his neighbours on Manar Park station in August 1929.[4] A little further afield, Paradise

Parrots were seen near Gin Gin in the 1930s. After that come only rumour and hope. Today, the Paradise Parrot has the tragic status of Extinct.[5] It's the only mainland Australian bird species known to have suffered that fate.

Jerrard's Paradise Parrot photos are historically momentous, not only as the first—and last—ever taken of the species, but also as the first photographs of an Australian bird to be accepted as confirmation of a rare species' existence. Before then, confirmation demanded taking a specimen by shooting the bird. That is still the procedure under certain circumstances today, but in the early 20th century, specimen-taking was rife. By the 1920s, birders' recourse to the gun was generating heated controversy, and the battle between supporters and opponents of private collecting was escalating steeply. The Paradise Parrot happened to be rediscovered at a moment of exceptionally acrimonious dispute over the rights and wrongs of converting living birds into dead skins. And the man who publicised the rediscovery happened to be an exceptionally vociferous participant on the anti-collecting side.

In 1922, seven proposals for the curtailment of private collecting were put before the council of the RAOU. The eminent South Australian ornithologist Edwin Ashby was outraged. These 'revolutionary resolutions', he thundered, failed to recognise 'the educative value of private collecting'. Collecting and skinning birds gave children an excellent introduction to ornithological studies, he insisted, as well as providing them with character-building experiences. Birds deserved proper protection, Ashby acknowledged, but curbing private collecting could only bring amateur contributions to the science of ornithology to an end.[6]

Chisholm's riposte was scathing. Paying lip-service to the proper deference of a young man to a 'veteran', he ridiculed Ashby's arguments, highlighting the absurdity of his simultaneous advocacy of protecting birds and skinning them. The 'private collector', Chisholm declared, 'is heavily handicapped as a public instructor or propagandist' for bird protection:

> Always his moralising is damned by the force of bad example. Chastity is the first essential of any salvationist.... The man who preaches private slaughter is on shifting sand as a public protectionist. His logic is bad; but even if it were good it would still be unconvincing to those whom science and scientists exist to serve, *viz.* the great mass of the people.

While conceding that museums should possess adequate collections of specimen skins, he contended that killing birds for personal collections sullied the moral standing of ornithology. The 'average private collector,' Chisholm declaimed, 'is a relic of barbarism and a perversion of civilisation. He is more; he is a relic of sin, masquerading under the honoured name of Science'.[7]

Ashby and Chisholm continued to clash through the 1920s and beyond: prominent participants in a bigger battle between pro- and anti-collecting factions that bedevilled birding in the interwar years. To some extent, it marked a generational divide, although that cleavage was far from clear-cut. Private collecting continued, albeit under constraint and controversy, but no one, as far as I know, attempted to convert a Gayndah Paradise Parrot into a skin. Although the displacement of the gun by the camera had not yet progressed far, Jerrard's photographs were accepted as conclusive evidence of the bird's existence.

Yet this was just the cusp of change, and rare birds were still not spared the collector's gun. To the contrary, old-school ornithologists insisted that authentication by a skin was exceptionally important in the case of rare species. Sight records were not enough. A.J.

Campbell in 1928 declared that 'except a specimen be procured, it is better not to record rare species merely from observations in the field'.[8] A year earlier, Ashby published an article on two *Neophema* parrots, the Orange-bellied and Scarlet-chested, which he described as among 'the rarest and least-known' of the genus, as well as 'among the most radiant birds on the Australian list'. Rare and radiant though they were, Ashby insisted that any new records of these species must be 'authenticated by a skin'.[9]

Campbell and Ashby had a point. Sight records *were* unreliable. With rudimentary optical equipment, cumbersome cameras, and embryonic field guides, having a bird in the hand was often the only way of guaranteeing correct identification. Paradise Parrots were certainly not the only species to be misidentified in the field, but after Jerrard's rediscovery came several sight records of these parrots that seem to have been based more on wishful thinking than skilful discernment. Florence Irby, for example, published some very dubious claimed sightings near her hometown of Casino in northern New South Wales in the late 1920s.[10] Arguably, overenthusiasm was also responsible for many later reported sightings.

In Australia, the validity of sight records remained a contentious issue through the 1920s and 1930s, entangled in contemporaneous controversies over the morality of killing birds to identify them. American controversies over sight records stretched back several decades earlier. There, by the 1920s, the supporters of sight records were in the ascendant, led by the inimitable Ludlow Griscom, the 'Dean of the Birdwatchers' as he was called, although he was also a professional ornithologist.[11] Their cause was furthered by the American innovator of the field guide, Roger Tory Peterson, who wryly remarked that 'the ornithologist of the old school seldom accepted a sight record unless it was made along the barrel of a shotgun'.[12] Peterson's own field guides did much to replace gun barrels with binocular lenses. Australia's lack of such sophisticated field guides as Peterson's was probably a factor retarding the acceptance of sight records here.

For the Paradise Parrot, there were photographs: more tangible and less contestable than mere sight records. However, photographs of rare species were seldom obtainable at a time when photographing a bird almost invariably entailed first finding its nest. Jerrard was fortunate that the rare species he rediscovered nested in termite mounds in open view. Indeed, by his own account, the mound provided a kind of stage on which the parrots posed: an ideal set-up for the cameras of the day. Few rare species were so obliging.

Its photograph may have been a factor protecting the Paradise Parrot from the collector's gun, but the real threats to its survival were more insidious. Jerrard knew what they were, explaining in 1924:

> The one undisguisable fact ... is that the advent of the white man has spelled destruction to one of the loveliest of the native birds of this country. Directly by our avarice and thoughtlessness, and indirectly by our disturbance of the balance so nicely preserved by nature, we are undoubtedly accountable for the tragedy of this bird.[13]

Although he was a grazier, he acknowledged that 'the most fatal change of all' had been wrought by the pastoral industry. Chisholm also nominated pastoralism, especially the associated burning of grasslands, along with trapping for the aviary trade and the ravages of feral cats, as the major factors. The Paradise Parrot, he lamented, 'has been brought to the very verge of extinction by human agency'.[14]

Jerrard and Chisholm may have identified the reasons for the Paradise Parrot's decline, but they were unable to do much about it. Chisholm wrote prolifically about the parrot, publicising its plight

Figure 12. Cyril Jerrard inspecting an abandoned Paradise Parrot nest in a termite mound, 1920s.

and pleading for its preservation. He ended his first newspaper piece on its rediscovery, published in *The Queenslander* on 12 August 1922, with a typically emotive plea on the birds' behalf:

> The next move in the history of the paradise parrots rests with the people of Queensland. It is for them to say if "the most beautiful parrot that exists" shall be wiped off the face of the earth. The governing authorities of the State have done their part—or at least done something—by according the lovely birds the total protection of the law; but this cannot be made effective unless Queenslanders are patriotic enough to say: "This slaughter of the Innocents has gone far enough."[15]

Like other birders of the day, Chisholm valued protective laws, but knew that they alone were insufficient to safeguard threatened species. Many believed that the most effective strategy was to stir the public conscience and cultivate a love of birds. That was Chisholm's forte.

The final chapter of his book *Mateship with Birds* was titled 'The Paradise Parrot Tragedy'. In it, he grieved that the 'extinction of a species is a ghastly thing' that was looming alarmingly close not only for this parrot, but for many of Australia's birds. He recounted the Paradise Parrot's near brush with annihilation as a cautionary tale to prompt his fellow Australians to end the needless destruction of nature. With a characteristic literary flourish, he concluded *Mateship* by urging readers to 'dispute the dangerous idea that a thing of beauty is a joy for ever in a cage or cabinet; and disdain, too, the lopsided belief that the moving finger of Civilisation must move on over the bodies of "the loveliest and the best" of Nature's children'.[16] When Chisholm wrote those words in 1922, he could not have known how deep the tragedy of the Paradise Parrot would run.

The pleas of Chisholm and fellow birders on behalf of the Paradise Parrot did not exactly fall on deaf ears, but they were inadequate to counter an ethos that privileged economic gain over avian loss. Besides, ornithologists then had a lamentably restricted repertoire of strategies to save endangered species. This was the case not only in Australia but overseas, too. American historian Thomas Dunlap recounts that in 1922—just a year after the Paradise Parrot was rediscovered—Whooping Cranes abandoned their last known nesting site in Saskatchewan, prompting fears that the species was, or soon would be, extinct. But, Dunlap observes, 'no one took action, not from indifference, but because no one knew what had to be done, or even how many cranes remained'.[17] Australia lagged far behind America in the knowledge and expertise needed to save threatened species. By the 1930s, some ecologically informed strategies for bird conservation were proposed, but such proposals were not implemented until well after the Second World War. By then, it was too late for the Paradise Parrot.

Yet few birders rushed to consign the species to extinction. For decades, many held hopes that the parrot still lived in remote corners of the land. Chisholm was one, clinging to a belief in the bird's probable survival until his death in 1977. He knew it was not a matter for dogmatism, telling his friend Jim Bravery in 1965 that 'the Paradise Parrot probably still exists ... but of course one can't be certain'.[18] Yet reported sightings kept coming sufficiently often to quell whatever doubts arose.

There was a spike in reported sightings in the 1960s and 1970s, perhaps a result of the increasing accessibility of the outback by four-wheel-drive vehicles and the increased popularity of birdwatching. A reported sighting in 1966 by a kangaroo shooter near Hebel in south-western Queensland drew an enthusiastic response from Chisholm. 'The species has been rediscovered', he excitedly told his friend Janey Marshall, 'apparently quite definitely'. He travelled out to Hebel to search for the parrot in company with field guide author Graham Pizzey. They found lots of Blue Bonnets, but no Paradise Parrot.[19] The kangaroo shooter's story proved illusory, soon joining the many others that fluttered on the edge of the imagination.

Yet reported sightings continued, as did expeditions to investigate them. One, in 1992, was led by Pat Comben, a keen birder who was then Queensland's minister for environment and heritage. The ten expert birders and National Parks officers who comprised this expedition combed an area of Queensland's Central Highlands where the Paradise Parrot had been reported shortly before. 'We were motivated by credible 1990 reports of the parrot being observed on a cattle station in the Dawson Valley', Comben recalled in 2021. 'After days of tough birding, we found no trace of the Paradise Parrot. On the final day, as I stood on an isolated sandstone bluff, uncertainty remained. I looked across the vastness of the largely unvisited area we were trying to cover. Not prime habitat, but big enough and isolated enough to hold tight its secrets.'[20]

There was a hint of hope in Comben's words, but he realised that the parrot's chances of survival are vanishingly small. Some birders think otherwise, maintaining faith that the Paradise Parrot lives on. Some still search for it. Optimism is admirable, but it is sadly prudent to accept the verdict of extinct.

Only one living person has seen live Paradise Parrots. He is Eric Zillmann, now aged 101 and living in Bundaberg. He last saw the parrots on his parents' property near Gin Gin in 1938 when he was only 15. But the memory is still vivid. 'I can see the bird now as clearly today as back then', he told renowned birder Greg Roberts in 2011. Eric and his father regularly saw a pair of the parrots when out mustering cattle in the 1930s, but thought little of it at the time because, as he put it, 'we did not know that the parrots were so rare'. In later life, he became a celebrated naturalist and legendary birder, but childhood encounters with Paradise Parrots are among his most cherished memories. Reflecting on them, Eric said 'I am humbled by what I regard as the most uplifting experience of my life.'[21] For the rest of us, who have never seen a Paradise Parrot and never will, remembrance of its tragic passing may also be humbling.

CHAPTER NINE

Noisy Scrub-bird: Hope

LIKE THE PARADISE PARROT, the Noisy Scrub-bird was first made known to science by John Gilbert, who collected a specimen on 3 November 1842 near Drakesbrook on the Darling Escarpment in Western Australia. He found it much harder to shoot than the parrot, because whereas the latter inhabited open woodlands, the scrub-bird skulks in tangled undergrowth. But Gilbert was a good shot, so a specimen found its way into the hands of John Gould, who named the species *Atrichia clamosus*. The generic name has since changed to *Atrichornis*, but the specific name survives, fittingly so since it translates as 'full of noise'. In Gilbert's description, the scrub-bird's call was 'so exceedingly loud and shrill, as to produce a ringing sensation in the ears. Precisely the effect produced when a shrill whistle is blown in a small room'.[1]

Like the Paradise Parrot, the Noisy Scrub-bird population plummeted after colonisation, probably also mainly because of changed fire regimes.[2] Twenty or so specimens were taken before 1889, when A.J. Campbell collected one at Torbay near Albany on the south-west coast. Thereafter, many sought the bird, none successfully. Sid Jackson hunted for it for five months in 1912–13 without success. His failure, after numerous others, led Campbell in 1920 to assert that the Noisy Scrub-bird 'is lost forever'.[3] Few birders expressed themselves with such finality. Even Gregory Mathews, who was renowned for supercilious certitude, allowed some leeway by designating the Noisy Scrub-bird 'apparently extinct'.[4] It was a verdict on which many birders agreed.

Many, on the other hand, maintained hope in the scrub-bird's survival. In his 1921 'Notes on the supposed "extinct" birds of the south-west corner of Western Australia', Edwin Ashby reported meeting a resident of Ellensbrook who had recently heard and seen the species. 'I feel confident that this bird still exists', Ashby averred.[5] When, in 1927, the RAOU held its annual campout at Normalup on the south-west coast, searching for the Noisy Scrub-bird was high on the agenda. 'Each morning a feeling of optimism, bred of eternal hope', James Pollard recorded in the *Emu* account of the campout; and 'each evening a sense of defeat, occasioned by empty collecting-bags'. 'Not that anyone believed it to be extinct', he hastily added; 'on the contrary, we came to the conclusion that it was very likely that the bird was there—but we might visit the locality a dozen times and never see or hear it'. This was 'grim country, able to hide its treasures', he explained, and 'we had no more explored the areas we had visited than ants and beetles and grasshoppers could explore it'.[6]

But the hoped-for sightings never came, so faith in the scrub-bird's survival dwindled with each passing year. In 1933, Hubert Whittell 'declined to believe' that the Noisy Scrub-bird was extinct.[7] Ten years later, in a review of ornithological knowledge of the species, he referred to the bird's 'extermination' as if that were a *fait accompli*, and made no attempt to question the growing assurance in its extinction.[8] Whittell himself had mounted several searches for the scrub-bird, none of which found any trace of it. His waning faith in the species' survival was shared by his colleague Dom Serventy. In their jointly written *Handbook of the Birds of Western Australia* (1948) they labelled the Noisy Scrub-bird 'Possibly extinct'. Two pages later, they referred to the 'probability' that no more information would ever be gathered on the species, implicitly because it no longer existed.[9]

The Noisy Scrub-bird was not the only species feared to be extinct. In the first half of the 20th century, a prodigious array of species was tagged 'extinct', though usually with a qualifier such as 'probably' or 'likely'. Only occasionally were early-20th-century ornithologists bold enough to pronounce on a species' extinction with finality, and even then their apparent assurance should not be taken at face value. In 1915, for example, A.J. Campbell declared that the Night Parrot 'has been exterminated,' and the following year proclaimed the Turquoise Parrot 'is now extinct'. Yet in the very same articles, he observed that it 'would be interesting to know' whether these and several other 'beautiful Australian Parrots still exist, or have been exterminated'.[10] Such inconsistencies flag the indeterminacies that hovered over avian extinctions. The readiness with which birders resorted to the language of extinction testifies to the intensity of their fears, but, with very few exceptions, the state of ornithological knowledge precluded certainty about whether those fears were well founded.

'Our knowledge is in many directions very scrappy', Edwin Ashby acknowledged in his presidential address to the 1926 RAOU congress, and 'we are still absolutely in the dark as to the extent and limitations of the range of habitat of many of our birds'. In this circumstance, he pointed out, it was impossible to ascertain whether a species was extinct or even, in many cases, under threat of extinction. Indeterminacy was all the greater for the fact that species consigned to probable extinction had a habit of reappearing. Ashby cited the recently rediscovered Paradise Parrot as just one example. Among others, he listed the Turquoise, Scarlet-chested, Orange-bellied, and Regent Parrots as species that had been considered probably extinct but had later reappeared, some in substantial numbers. Lack of reliable information on avian extinctions, he maintained, could be traced to one main reason: 'Australian ornithology, especially from the ecological side, is still in its beginnings. Much rarer than many of our supposedly extinct birds are trained, working field ornithologists.'[11]

Keith Hindwood, arguably Australia's greatest amateur ornithologist, made a similar point in 1939:

> Australia is such a vast continent that statements regarding the extinction, or near extinction, of any species should be made with due caution. Changes in climatic conditions, seasonal variations in food supply, and settlement influence the distribution of birds and often cause them to leave their known haunts. Only too often it is then assumed that they are on the verge of extinction. Twenty years ago this was said of some of the *Neophema* Parrots, and more recently of the Flock Pigeon (*Histriophaps histrionica*), but those species have been recorded in considerable numbers during the past few years. The absence of records signifies, more often than not, the absence of observers.[12]

A committed conservationist, Hindwood was not downplaying the threat of avian extinctions. He was making the point that scanty data counselled caution.

Cyril Jerrard agreed. In a 1926 *Emu* article, he quoted with approval another naturalist's statement that "'one begins to wonder whether there are not more of our so-called extinct birds still in existence if the many remote districts of the State were examined by similar careful observers'". He was writing not about the Paradise Parrot, but the Black-breasted Button-quail, several of which Jerrard had found on his Gayndah property. He called it a 'nearly extinct species', while the editor of the *Emu* designated it a 'bird which it was feared had become extinct'.[13] Having found on his own small grazing run two species widely feared to be extinct, Jerrard was right to suggest that asseverations of avian extinction were premature. Few of his birding contemporaries would have contested that.

Yet the circumstance of the Noisy Scrub-bird differed from those of the other mainland Australian species that were feared to be on, or over, the brink of extinction. The others were found to survive—sometimes soon after being consigned to probable extinction—or at least to have left some tangible trace of endurance. The latter was the case for the Night Parrot, Australia's most famous avian escapee from extinction.

The last Night Parrot specimen was collected by Frederick Andrews in the Gawler Ranges in the early 1870s, and the failure of numerous subsequent searches led to the species often being labelled extinct or probably extinct. Yet evidence of the Night Parrot's survival—sightings, glimpses, feathers, and bits of desiccated body—was found in every decade of the 20th century, and a great many birders, recreational and professional, refused to endorse the verdict of extinct.[14] The Noisy Scrub-bird lacked even scrappy evidence of continued existence, making hope hard to sustain.

In 1948, the RAOU and the Western Australian Historical Society collaborated to erect a memorial at Drakesbrook commemorating John Gilbert's discovery of the scrub-bird 106 years earlier. The suggested wording for the plaque, formulated by RAOU president Dom Serventy in May 1948, designated the Noisy Scrub-bird 'a remarkable bird, now extinct'.[15] The memorial erected on 12

September 1948 carried words closely following Serventy's in all other respects, but instead of 'now extinct' it described the species as 'a sweet-voiced bird of the scrub which has since been rarely seen'. I have been unable to ascertain why or at whose prompting the word change was made, but presumably some influential birders rejected the finality of 'now extinct'. The thread of hope was thinning, but had not yet broken.

Alec Chisholm was one who refused to allow that thread to break, for any Australian bird. Writing on the Noisy Scrub-bird in 1950, he stated that, 'It is possible, if not probable, that examples still exist in isolated areas, but if so they must be extremely rare.'[16] The following year, he published a two-part article on scrub-birds in the *Emu*, in which he maintained that:

> It is, of course, somewhat risky to refer to any bird of the Australian continent as extinct, for, although some have not been recorded for years, not a single mainland species is definitely known to have vanished since settlement began, and it may be that the retiring Noisy Scrub-bird will one day be rediscovered in some quiet corner of the south-west.[17]

Yet while he was not prepared to write off the scrub-bird, the tone of Chisholm's statements suggests he was less hopeful of its prospects than those of the Paradise Parrot. After all, he kept receiving reported sightings of the latter species, whereas on the Noisy Scrub-bird there was silence.

The very few claimed encounters with Noisy Scrub-birds got no publicity and led to no discoveries. In 1954, A. Kalmins of Katanning told Hugh Wilson, chairman of the RAOU Conservation Committee, that he had found the nests and eggs of the Noisy Scrub-bird. Excitedly, Wilson responded:

Although it is often stated to have become extinct, I have never believed that this was so, and in a report which I wrote to the International Conference on Bird Protection now being held at Scanfs, Switzerland, I stated that I believed the Noisy Scrub Bird would be found again. I did not expect that this prediction would be fulfilled so soon. Of course, one must make sure in these matters, but so far from disbelieving you I am acting on the assumption that your identification will be confirmed.

Wilson, who lived in Melbourne, asked the Perth-based CSIRO scientist John Calaby to investigate Kalmins' claim. Calaby discussed the matter with fellow Perth ornithologists Dom Serventy and Eric Sedgwick; he gave Wilson a sceptical reply.[18] Kalmins' identification was almost certainly erroneous, but the correspondence it generated illuminates the faith some birders maintained in the species' survival.

Their faith was well-founded. On Sunday 17 December 1961, schoolteacher and amateur naturalist Harley Webster was fishing off the beach at Two Peoples Bay when he heard birdcalls he had never heard before:[19]

The loudness and richness of the calls were remarkable and I began to hope that it was indeed that will-o-the-wisp that has lured ornithologists in the past seventy-odd years into the thick scrubs of the South West—*Atrichornis clamosus*.... I hardly caught a glimpse of it that day, it was so adept at keeping itself under cover. I came away in the evening with impressions of a brown bird with a call that really made my ears ring and with the knowledge that it was almost certainly the Noisy Scrub-bird.[20]

The following Saturday, he was back at the same place, and got better views. Still better views on 24 December assured Webster that it was indeed a Noisy Scrub-bird. He announced his rediscovery in the *West Australian* on Christmas Day 1961, and on 28 December Dom Serventy travelled down from Perth to confirm the identification.

That's a conventional version of the story of the rediscovery of the Noisy Scrub-bird. Research by Allan Burbidge and Alan Danks, conservation biologists who had major roles in recovery strategies for the species, reveals a more complicated sequence of events, entangled in the perennially controversial issue of collecting.[21] In this version, Charles Allen, an amateur egg collector, heard and saw Noisy Scrub-birds while searching for Red-eared Firetail nests at Two Peoples Bay on 5 November 1961. He notified ornithologist Julian Ford, who went to the area, failed to find scrub-birds, but did find Western Bristlebirds and shot three of them for specimens. One of these had been under observation by Webster, who complained of the shooting to the Western Australian Museum. In his letter of apology to Webster, Ford let slip that he had been at Two Peoples Bay to investigate a reported scrub-bird sighting. Webster, it seems, had been forewarned of the likely presence of the scrub-bird.

The story becomes still more complicated by the likelihood that Charles Allen had seen the Noisy Scrub-bird at Two Peoples Bay back in 1942. He told only a select few at the Western Australian Museum, one of whom, Ken Buller, travelled to the site in 1945, found no scrub-bird, but did collect a bristlebird. Lacking official confirmation, Allen failed to publicise his encounter with a scrub-bird, and attention was diverted to the successful acquisition of a Western Bristlebird, a species whose survival had been considered almost equally precarious. Allen's probable scrub-bird sighting of 1942 came to light only after Webster announced his own sighting of 1961. Webster seems to have accepted the veracity of Allen's claim, but relationships soured in early 1962 when he became convinced that Julian Ford was publicising Allen's earlier encounter in an attempt to deny Webster recognition as the rediscoverer the Noisy Scrub-bird. Evidently peeved, he vented his frustration to several birders, telling them that 'my juvenile friend J. Ford has succeeded in publishing a cunningly fabricated prior claim for C. Allen'.

Burbidge and Danks consider Allen's claimed 1942 encounter with the scrub-bird highly credible, and his November 1961 sighting certain. However, Allen failed to publicise either encounter, whereas Webster got his into the newspaper the day after he became certain of what he saw. Personality differences seem to have been involved, Allen being a diffident man who shunned the limelight, while Webster had a keen appreciation of public recognition. Putting it another way, Webster was the first to both identify a Noisy Scrub-bird and have the self-assurance to publicise his identification.

Ego appears to have been among Webster's motivators, as it has been in many episodes in the history of birding. But this does not detract from the either the magnitude of his achievement or the genuineness of his solicitude for the scrub-bird. His suspiciousness of Ford stemmed from his belief that the ornithologist was excessively eager to convert the scrub-bird into a skin. Though not totally opposed to specimen collecting, Webster considered Ford's priorities perverse in putting acquiring a specimen ahead of conserving the species. Whether or not his appraisal of Ford was accurate, Webster was deeply committed to the preservation of the scrub-bird.

Libby Robin has examined Webster's field notebooks, in which he agonised over the choices available to him. The notebooks reveal that as ardently as he sought recognition of his 're-discovery rights' he also strove to protect the bird from all threats, including collectors' guns. Robin goes on to locate Webster's rediscovery in the context of the times. The 1960s, she observes, 'saw acute acrimony between old-school amateurs and the emerging class of new professionals—and this friction nearly tore the RAOU apart within a few years of the Noisy Scrub-bird find'.[22] Disagreements over the Noisy Scrub-bird were no more than hiccups in that convulsion, but, like the Paradise Parrot, the scrub-bird happened to be rediscovered at a time of exceptionally intense conflict within the birding community. Perhaps there was never a time when birding was free of acrimony, but there were peak periods, and the 1960s was among them.

Yet while there were mounting tensions between amateurs and professionals, there was also continuing cooperation. Webster collaborated extensively with scientists, particularly with Western Australia's pre-eminent ornithologist, Dom Serventy. In January 1964, both men participated in the first-ever capture and banding of a Noisy Scrub-bird, in a team that also included Dom's brother Vince Serventy (a popular natural history writer and film maker), Graham Pizzey (journalist), Harry Shugg (Western Australia's chief warden of fauna), Robert Stranger (an enthusiastic birdwatcher and self-described 'drifter'), and Syd and Rica Erickson (acclaimed amateur naturalists). It's an array exemplary of the long tradition of amateur–professional collaboration in birding.

In any event, the threat to the scrub-birds came not from shotgun-wielding ornithologists, but from real estate developers. Two Peoples Bay was the site of a proposed town, to be named Casuarina, whose construction would have destroyed, or at least degraded, the only known habitat of the Noisy Scrub-bird. Conservationists, Harley Webster prominent among them, rallied to the cause, with support from persons as eminent as HRH Prince Philip, the Duke of Edinburgh. Their campaigning was successful, Dom Serventy reporting in June 1966 that the planned town 'has been definitely scrubbed' and the scrub-bird's habitat protected in a nature reserve.[23] With a tiny population in a minuscule area, the Noisy Scrub-bird was still vulnerable, but the efforts of ecologists, ornithologists, conservation biologists, and volunteer birders ensured that the species not only survived, but increased. Successful translocation programs further boosted its numbers. Today, the IUCN Red List ranks the Noisy Scrub-bird as Endangered.[24] It's a far preferable status to what it was feared to be for much of the 20th century.

Throughout that time, a dwindling few clung to hope in the scrub-bird's survival. That they were right attests to the wisdom of maintaining hope in the face of adversity. This might seem at odds with my remark in the preceding chapter, that it is 'sadly prudent' to accept the verdict of extinct in the case of the Paradise Parrot. Prudence is, I think, appropriate in that instance, so that scarce conservation dollars and energies aren't wasted on a lost cause. But prudence isn't synonymous with certitude. For practical purposes, it's wise to accept that certain species are extinct, but that doesn't mean we must abandon all hope. 'Hope springs eternal in the human breast', the poet Alexander Pope famously declaimed. May birds such as the Noisy Scrub-bird continue to justify Pope's aphorism.

CHAPTER TEN

Sarus Crane: Discovery

FIVE YEARS AFTER HARLEY Webster rediscovered the Noisy Scrub-bird, another avian species was added to the Australian list. This one was not lost and later found. Its presence in Australian was hitherto unknown to science. The discovery began on the morning of 13 October 1966, when three birdwatchers realised that the six Brolga-like birds they were observing on the floodplains near Normanton were not Brolgas at all. The next day, they saw two more of the 'big cranes with red upper-necks' at Lake Woods near Burketown. It took the trio—Billie Gill, Fred Smith, and Eric Zillmann—some time to figure out what they were.

According to Smith, his first inkling that they may have been Sarus Cranes came on Boxing Day 1966 when, idly browsing through *Collins Pocket Guide to British Birds*, he stumbled upon an illustration of that species. The then-known range of the Sarus Crane was south and south-east Asia, but they appeared in the book because they occasionally escaped from British zoos. Suspicions aroused, on 28 December Smith examined the single representative of the species held by the Melbourne Zoo.[1] His zoo visit provided, in Billie Gill's words, the 'final clincher' in identifying the bird that she, he, and Zillmann had seen near Normanton as a Sarus Crane.[2]

Identification clinched, Gill was keen to return. A resident of Innisfail, she lived closer to Normanton than either Smith or Zillmann. Among other things, she wanted to photograph the cranes, for when she had first seen them, both her companions 'had cameras and 400mm lens', but they 'never thought of photographing the birds at the time'. Like Smith and Zillmann, Gill was an amateur birder, with immense enthusiasm and expertise, but little money and limited time to spare from the pressing demands of earning a living and raising her eight children. She especially regretted her lack of adequate means of transport to take her back to the Gulf, telling CSIRO ecologist Francis Ratcliffe in January 1967: 'oh how I wish I had some money so I could buy a good long wheel based Land rover. I think they are the most wonderful of motors.'[3]

Gill had bigger worries than her lack of a Land Rover. She feared what scientists might do once they knew that Sarus Cranes were conveniently accessible in northern Australia. Ratcliffe was a scientist she trusted and in whom she confided, pouring out her concerns in several letters in early 1967. 'Now our problem is this', she told him, 'How to we [sic] be assured that miserable museum collectors and other murderers will not go at once up there and shoot the lot?' She was keen to publish the find, but concern for the cranes' safety held her back.[4] Birdwatchers in the 1960s could be deeply distrustful of scientists' collecting zeal.

Ratcliffe replied that he could 'understand that you may feel a bit hesitant to publish and possibly invite the attention of collectors', but suggested she was being 'a trifle over-anxious'. After discussions with Warren Hitchcock at the CSIRO wildlife division, he reassured Gill that 'the danger from museum collectors was likely to be negligible'. Scientific collecting was strictly regulated, he reminded her, at the same time admitting that 'it might be necessary for one [specimen] to be taken just to put the identification beyond doubt'.[5] Gill was more or less mollified by this, conceding that, 'I guess one skin is alright ... so long as all of them are not wiped out.'[6]

Eventually, Gill made it back to Normanton on 25 April 1967, in company with her daughter Jean Gill, P.L. Duve, and Bruce Cook. They found 15 Sarus Cranes, including two pairs with flightless young, confirming that the species bred there.[7] During that visit, Cook took the first photographs of the species in Australia. Shortly afterwards, on 29 July, he identified 20 Sarus Cranes among about 40 Brolgas at Willett's Swamp on the Atherton Tablelands. He promptly told local birder Jim Bravery, who saw the mixed flock of cranes at the swamp the next day.[8] These were the first records for the Atherton Tablelands.

Before long, Sarus Cranes were found to be quite common on the Tablelands, as well as in the Gulf Country and Cape York Peninsula. In 1970, Gill reported having seen many in those areas over the preceding three years, and even a breeding pair at Cattle Creek near Ingham.[9] By 1990, when Fred Smith was publishing a regular bird-identification column in the *Bird Observer*, his piece on the Brolga and Sarus Crane stated that the latter 'occurs over a wide range of the Australian north ... and, in some places, is present in large numbers'. He added that, 'Although the Brolga and Sarus Crane are generally similar in appearance, they are relatively easy to tell apart.'[10]

Having seen both species, sometimes close together, I can vouch for the accuracy of that statement. Both are stately silver-grey birds, but the Sarus Crane has bare red skin extending about a third of the way down the neck, no dewlap, and pink legs, while the Brolga's red skin does not extend beyond the head; it has a prominent dewlap and dark-grey legs. I knew those distinguishing features before I saw the birds, having read about them in field guides. Field guides aren't only for use in the field; they enable birders to do their homework, so they come to the field already primed to identify what they expect to find. Lacking that benefit, the puzzlement of the trio on the Normanton floodplain in 1966 is understandable. Yet the relative ease of distinguishing Sarus Cranes from Brolgas raises the question of why it took so long for birders to do so.

One possibility, mooted since the birds were first seen in Australia, is that Sarus Cranes are recent arrivals.[11] However, scientific studies soon suggested otherwise. In 1988, avian systematist Richard Schodde classified the Australian birds as a distinct subspecies, noting that the 'presence of an endemic subspecies of the Sarus Crane in Australia implies that it has been present in north-east Queensland since long before the period of European colonization'.[12] He named the subspecies *Grus antigone gillae* after one of its discoverers, Billie Gill.

Schodde's assurance notwithstanding, some birders continued to assume that the Sarus Crane's recent discovery meant it was a recent arrival. Twitcher Sue Taylor made a remark along those lines in 2001, with an implication that its presence was both a consequence and a cause of ecological disruption.[13] Ornithologist Danny Rogers promptly chastised her for casting Sarus Cranes 'as foreign birds that have introduced themselves into Australia in historical times and started to oust the native Brolga'. 'Australian Sarus Cranes are morphologically distinctive', Rogers insisted. 'They are a natural part of the Australian avifauna and it does them no good to suggest otherwise.'[14]

But the question would not go away. In late 2004, John Grant, an Atherton Tablelands resident crane researcher, published an article arguing that the Sarus Crane's length of residence in Australia was still an open question. Fred Smith retorted that as a recognised subspecies, the crane 'must have been here a very long time'. Grant was unconvinced, citing genetic and DNA evidence that kept open the possibility of the cranes having recently arrived from Asia.[15] A 2014 article co-authored by Grant, Gill, and several other birders maintained that possibility, stating that 'the jury is still out on how long the Crane has been in Australia'.[16]

The jury seems now to be reaching a verdict: the species has been in Australia for a very long time. Recent studies by Tim Nevard and colleagues show that the Australian subspecies of Sarus Crane is genetically disjunct from Asian subspecies, reinforcing Schodde's assertion that the bird must be a long-term Australian resident.[17]

Added evidence comes from Indigenous knowledge of the bird. John von Sturmer, an anthropologist who worked with various Aboriginal groups in the western Cape York Peninsula, found linguistic and cultural evidence of long familiarity with the species.[18] According to anthropologist and linguist Peter Sutton, in the Wik-Ngathan language of the western Cape York Peninsula, the Sarus Crane is 'yoompenham', while the Brolga is 'thuulk' or 'korr'', the latter possibly a borrowing from the neighbouring Wik-Mungkan language.[19] Both von Sturmer and Sutton remark that western Cape York Peninsula Aboriginal people consider the Sarus Crane better eating than the Brolga, and if opportunity offers they will selectively hunt the former.

However, Aboriginal peoples' awareness of the Sarus Crane leads into further mysteries. The crane's range in northern Australia was not terra incognita for earlier generations of ornithologists, and those who went there routinely consulted local Aboriginal people on avian matters. For one of those early ornithologists, Donald Thomson, consultations with Aboriginal people were more than routine. His Cape York Peninsula expeditions of 1928 and

1932–33 were mounted expressly to carry out ornithological and ethnographic research simultaneously. It was as an anthropologist and champion of Aboriginal rights that he later won acclaim. On Cape York Peninsula, he was in constant communication with local Aboriginal people and acquired at least a working knowledge of local languages. Yet he made no mention of any crane other than the Brolga.[20] The Aboriginal people with whom he interacted every day were well aware of his ornithological interests, yet they failed to tell him about the Sarus Crane. Perhaps Thomson didn't ask the right questions.

There's anecdotal evidence of folk knowledge of the Sarus Crane among the local settler population, too. Gordon Beruldsen reported speaking with '"old-timers" at Normanton', who told stories indicating that they knew about 'the Red-legged Brolga' in the 1920s.[21] So, in all likelihood, the presence of a variant crane (later identified as Sarus Crane) was known to locals, Indigenous and non-Indigenous alike, although they may not have categorised the variant according to the scientific concept of a species. However, that knowledge failed to be transmitted to ornithological circles. Why it took so long for birders to notice such a big, bold bird as a Sarus Crane remains a puzzle.

The Sarus Crane was not the only new species added to the Australian list in the 1960s. Most additions were vagrant waders or seabirds, some of which have since been recognised as rare but regular visitors. Fred Smith had a special knack for finding them, with three to his credit before the Sarus Crane. In March 1962, he reported the first Buff-breasted Sandpiper in Australia, at Altona near Melbourne; on 22 December the same year, he found Australia's first Red-necked Phalarope at Werribee Sewage Works; and on 6 February 1966, he recorded the first Wilson's Phalarope at Lake Murdeduke in western Victoria.[22] He kept finding more thereafter, to the extent that, according to a *Wingspan* article of 2009, 'for a while, finding a new Australian bird became known as "doing a Fred Smith"'.[23]

That issue of *Wingspan* was devoted to 'the joy of discovery that watching birds can bring'. Whether it comprised finding new species or finding new facts about well-known species, editor Sean Dooley explained, 'being involved with the natural world, and birds in particular, means there is always something new to discover'. He went so far as to proclaim that 'one of the joys of birdwatching is that it is one area in life where the potential to make discoveries is boundless'.[24] Peter Slater went further, claiming that, 'Every birdwatcher, I am sure, dreams of adding a new bird to the Australian list, whether it is a migrant/vagrant from elsewhere, or a totally new species.' That's going too far. I'm sure the rare birders who add to the national list get a buzz out of it, but I doubt that most birders entertain such dreams. Birdwatchers love to discover things, but most of them, most of the time, are content with more modest discoveries: an out-of-season Koel, for example, or a range extension of the Freckled Duck. Slater was right, however, to warn that with discovery came responsibilities, and discovering a species new to science entailed 'an unpalatable aspect, namely the taking of specimens'.[25]

Like most birds discovered in this country since the turn of the 20th century, the Sarus Crane was not new to science, but merely new to the Australian list. In the decade of its discovery, however, two Australian endemics hitherto unknown to science were found in remote parts of the interior. On 7 July 1967, Norman Favaloro, a small-town solicitor, amateur birder, and renowned egg-collector, discovered the Grey Grasswren in the Bulloo Overflow in north-western New South Wales.[26] Favaloro himself collected the type specimen not only of the bird, but of its nest and eggs as well, apparently without provoking controversy. Things were very different for the other new species of the 1960s, Hall's Babbler. It was discovered not by an amateur, but by professional ornithologists participating in the Harold Hall British Museum expeditions of 1962–68. Unlike the Grey Grasswren, the discovery of Hall's Babbler was mired in controversy over collecting from the outset.

When news of the forthcoming Hall expeditions broke in 1962, senior members of the Bird Observers Club voiced strident opposition, seeking to stop the slaughter they thought was about to be unleashed upon the birds. They orchestrated a letters-to-the-editor campaign involving metropolitan and provincial newspapers throughout the country. They made representations to prime minister Menzies. Throughout the middle months of 1962, every issue of the *Bird Observer* carried a supplement keeping members informed—and agitated—about the expedition and its anticipated bloodletting. Among other things, they contended that modern technologies, including photography, rendered collecting redundant. The August supplement, headlined 'Keep Fighting!', warned that, 'The British Museum (Natural History) Collecting Expedition, together with several Australian Museums, are now awaiting to move in for the kill.' All supplements fanned fears that collecting might sound the death-knell for vulnerable species.[27]

The bird observers did not maintain their rage. When, over a decade later, *Birds of the Harold Hall Expedition, 1962–70* was published, it got a glowing review in the *Bird Observer*. 'The expedition yielded a rich harvest of scientific knowledge', the reviewer enthused, with Hall's Babbler discovered, several little-known species 'virtually rediscovered, and many other rare species closely studied'. 'Initial hostility' to the expedition, the reviewer guardedly remarked, 'was mollified when assurances were given that there would be no indiscriminate collecting'.[28]

But collecting continued to rankle. When, in the March 1977 issue of the *Bird Observer*, Lawrence Courtney-Haines remarked that, 'I have not as yet had the privilege of converting [a Reed-Warbler] into a cabinet study skin', it drew a caustic retort from fellow bird observer Gordon Clarke. It 'is beyond my comprehension', Clarke fulminated, how anyone could conceive killing a glorious songbird as a 'privilege.' But his retort exposed something far different from the bird-skinner's presumed callousness.

Courtney-Haines replied that he had 'never shot at, killed or harmed a bird in my life' and was 'utterly against collecting birds for any purpose whatever, including the collecting of birds for museums'. All his taxidermic work, he explained, had been on either aviary fatalities or wild birds found dead after some mishap, such as colliding with a window. More than that, he regarded birds 'as the greatest of God's art-forms', and he practised 'the romantic and gracious art of bird taxidermy' upon them as an act of reverence.[29] Clarke could hardly have imagined how mistaken he was when he castigated Courtney-Haines as a killer, but their interchange highlights the extreme delicacy of birders' sensitivities about collecting.

Those delicacies were put on display many times thereafter. In March 1993, Dr Richard Thomas published a letter in *Wingspan* that began by defending twitching as a viable form of birding, then slid into a broadside against collecting, describing it as a sordid procedure whereby a bird is 'murdered by a so-called scientist to add to his morbid collection'. Inevitably, museum ornithologists responded. Les Christidis took up the cudgels on their behalf, explaining scientists' continuing need for specimens and decrying the 'misinformation and hysterical language' deployed by Dr Thomas. Their vitriolic exchange in *Wingspan* continued for a year, neither participant showing the least sign of giving ground.[30]

While birders disputed the rights and wrongs of collecting, several Sarus Cranes were added to specimen drawers at the Australian National Wildlife Collection and the Queensland Museum. Although Ratcliffe had assured Billie Gill that it 'might be necessary' to take just one specimen, in fact six were taken: the holotype, taken near Karumba on 18 August 1984, plus five paratypes. Gill herself prepared the type specimen, having moved from Innisfail to Canberra in 1975 to take employment at the Australian National Wildlife Collection.[31] As a curatorial assistant there, the horror of specimen-collecting she had expressed in 1967 presumably had mellowed.

Regardless of the few sacrificed to science and the on-going disputes over that process, Sarus Cranes prospered in Australia. Probably, they benefitted from environmental transformations such as those on the Atherton Tablelands, where dense forest, a habitat unsuited to the species, has been converted into cultivated fields.[32] Whether there has been an actual increase in number is uncertain. The only reliable counts have been on the Tablelands, where the cranes overwinter, but do not breed. There, over 3,000 individuals have been recorded in a single season, but the numbers fluctuate, and the proportion of the total Australian Sarus population that overwinters on the Tablelands is unknown.[33] Still, over the last few decades we've acquired a great deal of knowledge about the ecology, distribution, numbers, and behaviour of Sarus Cranes in Australia.[34]

Much of this knowledge comes from the efforts of recreational birders. In 1997, Malanda resident Elinor Scambler initiated an annual Crane Count on the Atherton Tablelands, coordinated by the Birds Australia NQ Group (now BirdLife Northern Queensland) and conducted by volunteers. They bring their own binoculars and telescopes—and expertise—to the roosting sites, where they record data on both Brolgas and Sarus Cranes. Under several changes of leadership, Crane Counts have continued to the present day, blossoming into a major citizen-science project that uncovers vital information about the cranes and their conservation needs, both local and global. According to the Ozcranes website, 'The Crane Count has become a highlight of the year where people connect with each other, with cranes and the environment.'[35] It's a research project that leads to discoveries of the kind most birders treasure most highly: not the grandiosity of adding a new species to the national list, but the quiet achievement of adding new information that will boost birds' prospects of survival.

CHAPTER ELEVEN

Rock Warbler: Belonging

'IN THE ROCK-WARBLER (*Origma rubricata*) New South Wales possesses a very remarkable little bird.' With those unassuming words, 22-year-old Keith Hindwood began his first substantive article, published in the *Emu* in 1926.[1] It was certainly not his last. By the time of his death in 1971, Hindwood had contributed more pages to the *Emu* than any other writer before or since. [2] He had also published hundreds of articles in other journals plus six books, and had established an international reputation as an authority on Australia's avifauna, respected by such ornithological luminaries as Professor Ernst Mayr of Harvard University. Four years after his initial 'Rock-Warbler' article, he was appointed honorary ornithologist at the Australian Museum, a position he retained for the rest of his life. He was a fellow of the RAOU and its president from 1944 to 1946; a fellow and sometime president of the Royal Zoological Society of New South Wales; a corresponding fellow of the American Ornithologists' Union from 1938; and a life member of the Gould League of Bird Lovers of New South Wales from 1931. In

1959, he was awarded the Australian Natural History Medallion for his outstanding contributions to ornithology.

Yet Hindwood had no academic qualifications and no formal training in ornithology. He was not a professional ornithologist, but earned his living from a printing and stationery business in Sydney. His prodigious ornithological output was entirely a labour of love. It was in that sense that this colossus of Australian ornithology was lauded by friends and colleagues as an 'amateur'. Hindwood was exceptional for the magnitude of his scientific achievements, but there's nothing unusual about amateurs contributing to ornithological science. They've done so since birdwatching's beginnings, and still do today, despite the proliferation of professionalisation and credentialism.

When Hindwood published his first article on the Rock Warbler, and for decades afterwards, almost all contributors to the *Emu* were amateur birders. Of the 21 articles in the issue in which his article appeared, only one was by a professional ornithologist. Among the other amateur contributions was another article on the Rock Warbler by another young member of the RAOU, Norman Chaffer, who later won acclaim as one of Australia's greatest bird photographers. In a style and tone common in the amateur-dominated *Emu*, Chaffer's 'Photographing the Rock-Warbler' lauded its subject as 'a very endearing little bird, one that attracts because of its novelty, its brightness, its picturesque habitat, its restricted range, and its unusual nesting habits'.[3]

The final phrase refers to the birds' practice of suspending their nests from the roofs of caves and rock overhangs, often choosing sites near, and even behind, waterfalls. From this, they earned two of their alternative vernacular names, Cave Bird and Cataract Bird, both now obsolete though arguably more descriptively apt than 'Rock Warbler'. Having adapted reasonably well to European settlement, they took to nesting in sheds, culverts, mineshafts, verandahs, and other constructions that resemble their former nesting sites. Hindwood reported on a pair that built their nest above the dynamo in the engine shed at the Jenolan Caves, and another that suspended their nest from a whitewash brush in the carpenter's shop at the same place.[4] Another enterprising pair attached their nest to the wire mattress under the bed of the superintendent of Sydney's National Park. They successfully raised three young despite the bed above being used nightly by the house's human occupants.[5]

Equally noteworthy is the bird's restricted range. New South Wales' only endemic avian species, the Rock Warbler is confined to a small sector of the central-eastern portion of that state. True to its name, it lives among rocks, but only particular kinds of rocks. The Rock Warbler is a bird of the sandstone gorges, gullies, and plateaux of the Sydney-Hawkesbury region, sometimes extending into adjacent limestone areas, but never beyond. It's not a common bird, except perhaps in a few favoured locales. Seemingly never keeping still, it flits restlessly over rocky outcrops, and creeps into crevices seeking insects and other small prey, often flicking its tail sideways, but seldom flying and then never far. It calls loudly and frequently, sometimes mimicking the songs of other species, more often shrilling its own melancholy notes among the sandstone.

The Rock Warbler was Hindwood's special bird. He began studying the species as a boy, and continued to devote it close attention throughout his life. Many birders have a special bird—a species of which they are particularly fond and around which cluster personal memories, associations, and emotions. For Hindwood, the Rock Warbler was an exceptionally fitting special species, since he and it shared the same geographical range—he carried out almost all his bird studies in the same restricted territory as that occupied by the Rock Warbler. His high repute as an ornithologist was built on the foundation of birding trips that seldom strayed beyond the Sydney region and adjacent Blue Mountains. Like many birders of his generation, he 'felt it was better to know one area intimately, than to spread his energies too far afield'.[6] Hindwood abided by that dictum more resolutely than most.

Born at Willoughby on 3 July 1904, he lived in and near that Sydney suburb all his life. His attachment to Sydney was so intense that, according to his friend Alec Chisholm:

> [W]hen I told him in 1932 that I was transferring from the Sydney *Daily Telegraph* to the Melbourne *Argus*, he gasped in amazement.

Figure 13. Norman Chaffer's Rock Warbler frontispiece for Hindwood and McGill's *The Birds of Sydney*, 1958.

"Dammit!" he said, "you can't do that. You've written a book on the natural history of Sydney, and so you're bound to stay here!"

Chisholm added that someone had called Hindwood 'the man with the sandstone mind', which he took as a compliment.[7] Like the Rock Warbler, he was bonded to the sandstone-strewn environs of the Sydney-Hawkesbury region, making them both true denizens of Sydney.

Fittingly, *The Birds of Sydney* is Hindwood's best-known book, written collaboratively with his friend and fellow Sydneysider Arnold McGill. Equally fittingly, its frontispiece is a photograph of a Rock Warbler, wings outspread as it flies to its nest, taken by another friend and Sydneysider, Norman Chaffer. Part field guide, part natural history study, the book describes the various habitats of the Sydney region, relating them to the underlying geology and the overlaying avifauna. It lists 392 species, pointing out distinctive field marks, and giving advice on when and where each species can most likely be found. It's a guide to birds; but equally it's a guide to Sydney in which birds (rather than, say, beaches and nightspots) are listed and described.

Birdwatching is inevitably about place and the sensory experience of place. It's a spatial practice in which birders must be attuned to the nature of particular places if they are to have reasonable prospects of seeing, hearing, and identifying the birdlife. They don't need to become experts in all aspects of a place, but they do have to heed the local environment, whether it be remote bushland or, as was often the case for Hindwood, urban and near-urban locales. Typically, birders build up mental maps or personal geographies of the places in which they most often pursue their pastime. Even those who wander far and wide in search of rare and outlandish species usually have a local patch (or patches) with which they are intimately acquainted and whose birds hold the fondness of familiarity. Birding connects us with place, just as it connects us with nature, for the nature to which it bonds us is not the abstract nature of the scientific text or the political tract, but the embodied nature of birds and animals, grass and trees, soil and water, in tangible tracts of territory.

For his attachment to a particular place, Hindwood was compared with the celebrated late-18th-century English parson-naturalist Gilbert White, whose *Natural History of Selborne* has become a classic.[8] And he was compared for more than his focus on a specific place, an item in *Birds* magazine noting also that, 'following the Gilbert White tradition, [Hindwood] was one of the great amateurs who have enriched the history of science'.[9]

Amateurs still contribute to science, now under the banner of 'citizen science'. Yet while there are continuities, the amateur naturalist tradition followed by Hindwood and his contemporaries differed in key respects from today's citizen science. Projects of the latter kind typically entail teams of volunteer data-gatherers working under the direction of professional scientists. Amateurs are the foot-soldiers; scientists, the generals. Accredited professionals analyse and interpret the data garnered by laypersons, with the former usually attributed authorship of publications while volunteers get, at best, a mention in the acknowledgements. Hindwood, by contrast, worked in a tradition in which the entire process was in the hands of the amateur naturalist, from initiating the enterprise, through field work, to writing and publication.

Hindwood not only carried the amateur naturalist tradition to new heights; he did so with modesty and affability. Allen Keast, who went on to an illustrious career as professor of biology at Queen's University in Canada, fondly recalled his youthful engagements with the 'Sydney ornithological fraternity' in the 1930s and 1940s:

> Keith Hindwood was the great expert on the birds of Sydney, had read everthing [sic] ever written (and, I am sure, spoken in private conversation) about them. He maintained a formidable card catalogue on which every record was inserted: and in conversation could immediately recall every record ... He was the "final court of appeal" on every subject at the meetings. He was the ultimate gentleman and enthusiast.[10]

Not only at meetings, but also on field excursions, Hindwood was looked up to as the expert, but one who was never condescending to, or dismissive of, those with less skill and knowledge. 'No query on birds was too small or too silly', Roy Cooper wrote. 'His answers were always lucid and prompt.'[11]

Nicknamed 'Lofty' for his height—he stood almost two metres tall—Hindwood was certainly not lofty in manner. Renowned for his generosity and geniality, he had a down-to-earth manner and a penchant for Aussie colloquialisms. His sociability was natural and unaffected, but it was also inherent to the natural history tradition in which he worked. Natural history was practised as a craft in which knowledge and skills were transmitted to others more by word-of-mouth and personal interactions in the field than by books and formal instruction. For his sociability, he differed from his favourite bird, whose scientific name was *Origma rubricata* when he wrote his 1926 article on it, but was later changed to *Origma solitaria*, acknowledging the Rock Warbler's solitary habits. Hindwood's habits were gregarious.

His writing was in the unpretentious, jargon-free style characteristic of natural history. It was nonetheless expressive with flashes of imagination, as in this excerpt from a 1929 article on the Rock Warbler:

Occasionally a bird will hover in the air and secure an insect from beneath a jutting rock, also it will enter under low shelving rocks, where, because of the confined space, it is unable to hop; it then creeps about keeping its body low down in the attitude of a rodent. Often a bird will fly to the summit of a boulder, utter a few notes and then, suddenly, will seemingly fall backwards and, as though drawn by some magnetic force, will alight on the vertical face of the same or a nearby rock. One commonly sees *Origma* hopping over the upright faces of cliffs with the greatest ease.[12]

It was a writing style that combined empirical accuracy with descriptive clarity, often enlivened with passages of narrative, but eschewing explicit theory.

Also following precedent among natural history practitioners, Hindwood had a keen interest in history. He and many of his birding friends delved into the past as a means of enhancing their understanding of nature, Hindwood's special historical interest being in the naturalists and nature artists of early-colonial Australia. On those topics, he published numerous articles in such outlets as the *Royal Australian Historical Society Journal*, although much of his historical work appeared as insertions and asides in his ornithological writings. In an article on the Rock Warbler, for example, he interposed discussions of the bird's calls and its mode of movement with a quote from John Lewin's early-19th-century writings on the species, adding a lengthy footnote giving a bibliographic history of Lewin's *Birds of New South Wales* of 1813 and its predecessor *Birds of New Holland* of 1808.[13] Hindwood's paramount interest was in birds, but his fascination with human interactions with birds came not far behind.

As well as the natural history artists of the colonial era,

Hindwood followed those of his own times, especially those who specialised in birds. In 1966, he met William Cooper, and instantly recognised the artist's brilliance. It was Hindwood who launched Cooper on a career that took him to the first rank among Australia's bird artists. Their early collaboration produced the 1968 book *A Portfolio of Australian Birds*, with text by Hindwood and paintings by Cooper. This was Hindwood's last book, but not the end of his relationship with Cooper. Their warm friendship was commemorated six years after Hindwood's death at a memorial service in the Royal National Park for which both the invitation and

Figure 14. Self portrait of Keith Hindwood with a White-eared Honeyeater on his head, c. 1924.

the memorial plaque featured a painting by Cooper. It was of a Rock Warbler.[14]

Hindwood's own artistry found expression in photography. Peter Slater ranked him among 'The Masters' of Australian bird photography, explaining that he 'possessed a most enviable quality, and one exceedingly difficult to acquire: the ability to press the shutter release at precisely the right time, to show the subject in its most characteristic attitude'.[15] As the quintessential Sydney birder, much of Hindwood's bird photography was done in urban settings. Sometimes this was overt, as in his photo of a Magpie-Lark nesting on a telegraph pole at a busy tramway junction in North Sydney.[16] Often, the setting seemed more sylvan, since Sydney had—and has—large areas of bushland within it and nearby. Among such photos was an especially charming self-portrait taken at Middle Harbour, of Hindwood as a young man with a White-eared Honeyeater plucking hair from his head for its nest. This was the kind of intimate interactivity with birds that he and his birding companions treasured.

Although devoted to the birds of the Sydney region, Hindwood sometimes studied those further afield. After his marriage to Marjorie Goddard in October 1936, the newlyweds honeymooned on Lord Howe Island. It's a measure of his obsession with birds rather than of any want of romantic passion that he became fascinated by the island's distinctive avifauna, and immediately set about studying it—on his honeymoon. His long article on the topic was soon afterwards republished as a book, *The Birds of Lord Howe Island*. In 1960, and again in 1961, he sailed the Coral Sea as one of three ornithologists on the survey vessel HMAS *Gascoyne*. The outcome was a highly respected scientific report, *Birds of the South-west Coral Sea*, by Hindwood, Dom Serventy, and K. Keith.[17]

On an earlier excursion beyond Sydney, Hindwood was drawn

into the controversies over collecting that were tearing the birding world apart. At the 1935 RAOU campout at Marlo on the Gippsland coast, museum curator George Mack shot a nesting Scarlet Robin in full view of the campers, who for the preceding few days had been admiring the bird carrying out its parental duties. The killing was pointless and apparently deliberately provocative: an assertion of a collector's assumed right to deal death at will, rather than a genuine endeavour to collect a museum specimen. In protest, Hindwood led a walkout of the New South Wales contingent.[18] He was not opposed to collecting per se; he often used skins from the Australian Museum to illustrate his own talks. But he was disgusted by Mack's wanton act of bloodshed.

The ugly incident at Marlo earned the RAOU a lot of unaccustomed bad publicity, and continued to rankle among members for years. Later in life, Hindwood got involved in even bigger controversies within the RAOU, at the heart of which lay his own orientation to ornithology.

Since its inception, the RAOU had been dominated by the amateur naturalist tradition, but by the 1960s the pressures for change were so strong that they threatened to shatter the organisation. In her history of the union, Libby Robin warned that the '1968 Revolution' should not be seen as solely a clash between amateurs and professionals. Other factors were involved, but, as Robin's own account reveals, the biggest fracture line was between adherents to an amateur naturalist tradition and exponents of a professional, scientific model of ornithological research. Some RAOU members spanned the gap, among them Hindwood, the amateur ornithologist who held the respect of professional scientists. He found himself in the awkward position of go-between.[19]

Perhaps inevitably, since it was belatedly following global trends, the RAOU was comprehensively renovated in the late 1960s. The process was traumatic, but the organisation was modernised and professionalised, and its journal, the *Emu*, transformed into an overtly and exclusively scientific publication. Hindwood had wanted change, but was dismayed by what he saw as a cadre of scientists running roughshod over the birdwatching rank-and-file. He vented his displeasure in a letter to William Cooper:

> I have little interest now in the RAOU with the blasted Junta in charge. Arrogant academics, I feel, and all bloody up in the air with scientific jargon, graphs, sonagrams, etc. Mathematical ornithology. Why! the pleasure of birds is in watching them, studying their habits in the field, enjoying the surroundings, etc. but these days too many coves look upon the hobby as a dry as dust, detailed, exercise in theories and suppositions.[20]

Hindwood's outburst was more than an expression of frustration by an amateur birder who knew he was on the losing side in a battle with professionals. It was a lament for the divorce of science from birdwatching.

Hindwood's lament was not entirely misplaced, but nor was it squarely on target. The RAOU was transformed, but its leaders, guided by savvy strategists such as Dom Serventy, went out of their way to placate amateur members and maintain amateur involvement. Insofar as birdwatchers were alienated by the union's makeover, they transferred their allegiance to other birding organisations. In the wake of the RAOU's turmoils, the Bird Observers' Club, hitherto focussed on Victoria, began attracting more interstate members, while in Queensland, New South Wales, and Tasmania, existing birding bodies were revitalised and new ones established.[21] Before long, there was rapprochement between birdwatching amateurs and ornithological professionals, the two not merely coexisting, but continuing to cooperate and collaborate.

Hindwood didn't live long enough to witness those developments. He died of a cerebral haemorrhage on 18 March 1971 while birding with two friends at Wattamolla in the Royal National Park. The place and the activity were fitting, but he was aged only 66. He had retired the previous year and had plans for more writing, including a handbook on the birds of New South Wales, and to visit places in Australia he had never seen, such as Cape York Peninsula. In a touching obituary, Hindwood's friend, fellow amateur birder, and fellow Sydneysider Arnold McGill said that 'he stood as tall in the estimation of his ornithological associates as he did in physical stature'.[22]

Like other birders of his generation, Hindwood was not afraid to connect human and avian worlds. Though less inclined to anthropomorphise than some of his friends, he found moral worth in the study of birds. Concluding one of his many articles on Rock Warblers, Hindwood noted that, 'A study of their habits will give the naturalist a fuller understanding of life, for there is at least one philosophic moral we can well consider from this sprite of the sandstone, namely harmony with environment.'[23] That was written in 1929. They are sentiments that still resonate today.

CHAPTER TWELVE

Collared Sparrowhawk: Innovating

DISTINGUISHING A COLLARED SPARROWHAWK from a Brown Goshawk can challenge even the best birders. Both have the same plumage pattern: grey-brown above; a rufous collar beneath a grey head; and underparts finely barred rufous and white. Both have yellow legs and eyes. They live in the same woodland and forest habitats, where both hunt birds from ambush. Their calls are almost identical. So similar are these species that, several times when accompanied by birders far more skilled than I, we've seen one, and they've admitted uncertainty. Once, on Cape York Peninsula, I was with a mini-bus full of birders—13 in all, including two experienced guides—who spotted one of these raptors by the roadside. After ten minutes of scrutiny, we all decided it was a Collared Sparrowhawk, only to have that identification confuted when it took flight a moment later.

The species' similarity notwithstanding, today's birders would be surprised by a 1949 *Field Guide to the Hawks of Australia* by Herb Condon, curator of birds at the South Australian Museum. In it,

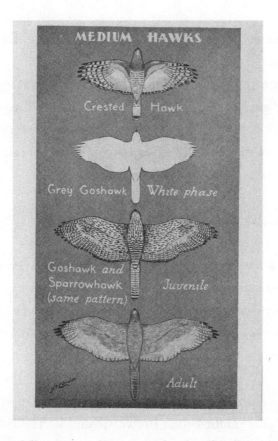

Figure 15. Shared illustration of Brown Goshawk and Collared Sparrowhawk in Condon's *Field Guide to the Hawks of Australia*, 1949.

every Australian raptor was illustrated, except that Brown Goshawk and Collared Sparrowhawk shared the same picture. The adjacent text stated that 'in the field it is almost impossible to distinguish the female [Sparrowhawk] from the male Goshawk', noting as the only difference that the 'Sparrowhawk has a longer middle toe than the Goshawk'.[1] The fourth edition of Condon's guide, published in 1966, was equally unhelpful in distinguishing between the two species. Goshawk and Sparrowhawk still shared the same picture, the descriptive text stating that 'the female [Sparrowhawk] cannot be distinguished from the male Goshawk in the field. Actual differences are:- Longer middle toe, weaker bill, and smoother tarsi in Sparrowhawk—visible only in museum skins'.[2]

Compare those concessions of defeat with the entries in a recent field guide, the CSIRO *Australian Bird Guide*. It gives multiple illustrations of both species (20 pictures in all), several explicitly pointing out distinguishing features. The text under Collared Sparrowhawk explains that this species is:

> most reliably distinguished by structural characters, especially *square-cut (not rounded)* tip to tail; more slender feet with *more elongated middle toe; brow-ridges smaller* than Brown Goshawk, *so head looks rounded with staring (not frowning) expression*. In soaring flight, *secondaries bulge beyond rest of trailing edge of wing*, most often seen in dashing powered flight, with more flicking wing beats and buoyant gliding than goshawk.

Further down, we're told that the voice of the sparrowhawk is 'slightly higher pitched than Brown Goshawk'. The entry for the latter species adds a few more fine points of difference.[3]

Partly, the contrast between Condon's advice and that in the CSIRO guide can be attributed to technology. Today's birders go into the field with binoculars and telescopes of an optical quality unknown to Herb Condon. Additionally, many carry cameras with powerful zooms, capable of capturing multiple sharp images that can be further magnified to uncover the finest of distinguishing features. With such optical equipment, today's field observer can see features that for Condon were 'visible only in museum skins'. Beyond optical technologies, there's the increasing sophistication of field guides themselves. Over the past 50 years, Australian birders have been treated to an efflorescence of field guides, and as the quality of successive guides has escalated, so have their users' expectations. We now take it for granted that species such as the Collared Sparrowhawk and Brown Goshawk are distinguishable in the field, if only we have the right cues.

Although Herb Condon did not make that assumption, his little raptor identification booklet of 1949 heralded a new era in Australian field guides. It was the first Australian field guide to adopt a style of illustration borrowed from the great American innovator of the genre, Roger Tory Peterson. Condon's hawks were depicted as Peterson's were: from a birder's perspective, in flight from below, with attentiveness to pattern rather than to details of plumage. (Actually, Peterson had borrowed this mode of depicting raptors from an earlier Canadian naturalist, Ernest Thompson Seton, but it was Peterson who popularised it and drew it into the mainstream of field guide illustration.[4])

In the 1950s and 1960s, several Australian field guides adopted Peterson-style modes of depicting and describing birds. All were guides to particular families and orders, including, as well as Condon's *Hawks*, his and Arnold McGill's 1952 *Field Guide to the Waders*. Guides to specific states and regions also showed evidence of innovation. Michael Sharland's 1945 *Tasmanian Birds: how to identify them*, for example, was the first Australian guide to make explicit use of 'diagnostic field marks' to distinguish species. By contrast, innovation sagged sadly in field guides of continental compass. There were only two, Leach's *Australian Bird Book* and Cayley's *What Bird Is That?*, and although they kept being reissued in reprints and new editions, they remained wedded to now-passé modes of depicting and describing birds. They showed none of the vibrancy and creativity that was invigorating field guides elsewhere, especially in America.[5]

Beneath the surface, however, innovation was bubbling in Australia. In 1955, CSIRO ornithologist Robert Carrick introduced Peter Slater to Peterson's field guides, an encounter that sparked

Slater's ambition to produce something similar for Australia.[6] He began serious work on the project in the early 1960s, in collaboration with Eric Lindgren, an ornithologist at the University of Western Australia. A little later, another duo embarked on compiling another Australian field guide in Peterson's style.

In 1965, publisher Billy Collins asked artist Robin Hill, then living in London, whether he was interested in producing a new Australian field guide. He was, and recommended Graham Pizzey to write the text.[7] Pizzey was keen on the project, but not on the terms that Collins offered—'they expect to get the job for next to nothing' he complained[8]—so it was mid-1966 before he signed a contract. Even then, the publishers offered to support the first year's work with an advance against royalties of only £750—'a piddling sum', Pizzey grumbled. Nonetheless, he took to the work enthusiastically, although he had to maintain himself and a young family by continuing his day job as a journalist.

Figure 16. Graham Pizzey (left) with artist Robin Hill, c. 1966.

By the time he embarked on the field guide in the mid-1960s, Pizzey had established a reputation as a talented nature writer. His talent was already apparent in his youth. In the 1940s, aged in his teens, he published several articles in Crosbie Morrison's *Wild Life* magazine, the second of which was on the Brown Goshawk. Pizzey described the bird in his characteristic visually evocative style:

> His back, wings, head and tail are a slatey grey with what seems a bloom of powdery grey over it; his breast is a mass of finely pencilled rufous bands. His eyes and legs are a lemon yellow and right around his neck is a superb rufous collar, making him the image of his smaller and much rarer cousin—the collared sparrowhawk.[9]

His descriptive flair would serve him well in his adult career as a field guide author. He was also, already, a skilled photographer, the cover of that issue of *Wild Life* carrying his photo of a Brown Goshawk at nest. He was not, however, a skilled artist, and his dependence on others for illustrations would later contribute to delays in completing his field guide. Hill ended his collaboration with Pizzey in 1969, to be replaced by a German artist, Hermann Heinzel, who in turn was replaced by Roy Doyle.

Among Pizzey's strongest supporters in the early years of his field guide project was CSIRO ecologist and founder of the Australian Conservation Foundation (ACF), Francis Ratcliffe, who believed that 'a field guide for bird identification is genuinely grass-roots conservation'.[10] Indeed, Ratcliffe maintained that a field guide 'is needed in Australia urgently' to help cultivate 'a public opinion supporting the ideas and practice of conservation'.[11] Considering Leach's and Cayley's books hopelessly outdated, he worked tirelessly to promote a Peterson-style guide. To that end, he sought financial support for Pizzey from the ACF's limited funds. The ACF executive agreed, but by the time they did so, in early 1967, they and Ratcliffe were aware that Slater and Lindgren were well advanced on their own field guide. Even-handedly, the ACF Executive Committee agreed to support both author teams with $1,000 for that year.[12] Slater and Pizzey had some differences of opinion, but on the importance of field guides as agents for conservation they were in earnest agreement.

Ratcliffe had earlier attempted to secure a Peterson-style field guide for Australia. Indeed, if his first efforts had gone smoothly, Australia would have acquired not merely a Peterson-style guide, but a guide by Peterson himself. In the early-to-mid-1950s, the CSIRO Division of Wildlife, headed by Ratcliffe, considered sponsoring the production of a field guide with Peterson as the proposed illustrator. This was around the time that Peterson published his first major non-American work, *The Field Guide to the Birds of Britain and Europe* (1954), in collaboration with Guy Mountfort and P.A.D. Hollom. Ratcliffe hoped he might do something similar for Australia, but was persuaded by his colleagues that the American was too busy and his services too expensive. A couple of years later, in 1956, Ratcliffe met Peterson in America and informed him of those discussions. Peterson told him 'that he would have loved to come to Australia and do the job' and 'was sorry that we did not approach him'.[13]

So Australia missed its first and best chance for a Peterson guide. When Peterson did, eventually, get to Australia in December 1965, he was hosted by Graham Pizzey, who had just begun work on his own field guide. Personal contact with Peterson, Pizzey reported, gave 'Robin Hill and I an excellent chance to discuss his approach to Field Guide production'.[14] After his departure in January 1966, Peterson continued communicating with Pizzey and offering advice,

Figure 17. International birders on Phillip Island, 1971. Left to right: Francisco Erize, Graham Pizzey, Barbara Peterson, Roger Tory Peterson.

in April that year sending him a copy of his latest field guide with the recommendation 'that we follow its layout closely'.[15]

Peterson visited Australia again in March and April 1971, when he was again hosted by Pizzey and taken birding by him around Melbourne. On travels further north, Peterson delighted in the spectacle of thousands of Rainbow Lorikeets being fed at the Currumbin sanctuary on the Gold Coast—a reminder that the greatest-ever field guide author was not above enjoying a bird show for tourists. Further north still, Peterson was astounded by his Innisfail host, Billie Gill, who habitually roamed the rainforests barefoot and barelegged. He and his wife, Barbara, 'asked tentatively, "Billie, do you see many venomous snakes?" She said: "No, I don't watch my feet"'.[16] Evidently, she was too engrossed in the birds.

Shortly after returning to America, Peterson told Pizzey how relieved he would be when he finished his Mexican field guide. 'Doing a field guide is rather like serving a prison sentence', he added.[17] This could hardly have been encouraging advice, since Pizzey had grown increasingly frustrated by how long his own guide was taking. The causes were many, but the primary one was Pizzey's perfectionism. As well as becoming conversant with the scientific literature on the birds in his book, Pizzey was determined to see them all in the field, to appreciate their jizz, their behaviour, their subtleties of plumage and demeanour. To that end, in 1967 he embarked on what he called an 'extensive caravan sortie': five months on the road with his family, travelling through vast expanses of the outback.[18] He did more trips with birding friends such as Claude Austin and Norman Wettenhall. For Pizzey, field familiarity was vital. By his own account, his breakup with Hill was precipitated by the artist 'producing plates quicker than he should' and failing to attend 'closely enough to the field feature of each species'.[19] The cost of Pizzey's contrary priorities was time.

Slater also took longer than anticipated to complete his guide, as he, too, strove to acquire field familiarity with as many birds as possible. In 1967, he moved from Perth to Innisfail in north Queensland, 'primarily to give him a chance of getting first-hand experience of the birds of that region'.[20] Around this time, suggestions were made that Slater and Pizzey might collaborate. That did not happen. Competition between them was friendly but palpable. In 1968, Slater's prospects were boosted by the addition of Harry Frith and a team of CSIRO scientists as contributors to his guide. The role of the new contributors, Slater told Lindgren, was to expedite completion of the guide and thus 'to get ours out first'.[21] A side effect was that when, in 1970, the first volume of the Slater field guide was published, Lindgren's name appeared as merely one contributing author among eight, rather than as equal author with Slater, as had long been envisaged.

Ratcliffe died in the same year as the first Slater volume (on non-passerines) was published, so he never saw the second (passerine) volume, which came out in 1974. Nor did he ever see the Pizzey guide whose production he had done so much to promote. It came out in 1980, to great acclaim from birders. A review in the *Bird Observer* began: 'This field guide seems to have been in the nest for an inordinately long period. Indeed many thought that it would never hatch. However, at long last it has fledged and a fine bird it is.'[22] A review by Alan Morris lauded it as 'the first true "field guide" of the birds of Australia'[23]—a bit extravagant, but indicative of the eagerness with which Australian birders welcomed the first single-volume guide in modern style and format.

After the 1970s, the floodgates opened. In 1986, two major field guides were published. Peter Slater collaborated with his wife, Pat, and son, Raoul, to produce *The Slater Field Guide to Australian Birds*. Very different from Slater's two-volume guide of the 1970s, this work remained the only compact field guide to Australia's birds until 2022. Not so compact was Ken Simpson's and Nicholas Day's 1986 *Field Guide to the Birds of Australia*. Essentially a scaled-down version of their 1984 book, *The Birds of Australia*, the 1986 guide incorporated the entire 70-page 'Handbook' from the earlier version, even though it was a compendium on avian natural history that offered little assistance in identifying a bird. The big innovation of both 1986 guides was to position descriptive texts and distribution maps on one page, and illustrations of those species on the facing page. This soon became standard format for field guides.

In the 1990s, Pizzey teamed with artist Frank Knight to produce not just a new edition of his 1980 guide, but a substantively new book. First published in 1997, its title, *The Graham Pizzey and Frank*

Knight Field Guide to the Birds of Australia, accredited the artist as co-creator of the guide. It quickly won admirers. When Lawrie Conole surveyed birders' field guide preferences in 1998, the Pizzey and Knight guide came equal first with the compact *Slater Field Guide*. Simpson and Day's book came well behind, with the Pizzey and Doyle guide lagging distantly fourth. In Conole's assessment, Pizzey and Knight's and the Slaters' guides were favoured by experienced birders, Simpson and Day's by novices, while Pizzey and Doyle's book was relegated to 'a few diehard fans'. Astutely, he noted that birders' field guide preferences were 'based on emotional factors as much as technical competencies'.[24] This was hardly surprising for a pastime pervaded with passion.

By the time that Conole conducted his survey, the field guide genre had further diversified. There were photographic guides, beginner guides, and a field guide to the nests and eggs of Australian birds, as well as a burgeoning output of guides to specific areas and to particular families and orders. In the latter category were several guides to raptors. There's something about birds of prey—their majesty, their lethal grace, their dominion over the heavens—that inspires awe and admiration, so their prominence in bird books is unsurprising.

Ornithologist Stephen Debus published *The Birds of Prey of Australia: a field guide to Australian raptors* in 1998. By this time, he could draw upon not only a wealth of birders' observations, but also a massive scientific compendium, the *Handbook of Australian, New Zealand and Antarctic Birds*, the relevant volume of which had been published in 1994. With this resource to hand, Debus could describe and depict in meticulous detail the distinguishing features of the Brown Goshawk and Collared Sparrowhawk. Fine differences of call were specified, and slight variances in soaring and gliding profiles illustrated; colour plates depicted both species in adult, juvenile, and immature plumage, as well as subspecies in the case of the goshawk.[25]

This all seems a long way from Herb Condon's 1949 *Field Guide to the Hawks*. Yet the similarities are stronger than the differences, and not only for the fact that both books aimed to help people identify hawks. One of the few points on which Condon shifted focus from identification was where he condemned killing raptors as 'foolish and tragic', since it upset 'the balance of Nature'.[26] Similarly for Debus's guide, though at greater length: the information it gave focussed closely on identification, except that each species' entry, as well as the discussion of raptors in general, had a section on 'Threats and Conservation'. It's a reminder that while field guides serve the practical purpose of helping us name the birds we see, they also serve grander objectives. That's why Francis Ratcliffe was so keen for Australia to get a modern field guide back in the 1950s and 1960s. He knew that field guides could be agents for conservation.

The two authors whom he helped to proceed to publication and who initiated the late-20th-century innovation of the Australian field guide, Peter Slater and Graham Pizzey, knew it, too. Extolling the necessity of a conservation consciousness in a 1995 interview with Gregg Borschmann, Pizzey avowed that 'the natural world is the absolute fundamental base, it's all our past, it's all our future ... the natural world is the great truth, the one thing that we need to know about. I personally believe that it can answer most of our spiritual cravings'. To which Borschmann responded: 'in that sense then, ... this field guide was your hymn, it was your testament'. 'Yes', said Pizzey.[27]

The words of that exchange call to mind British birder Simon Barnes' declaration that 'every field guide that was ever printed is not merely a book of helpful hints on how to tell one bird from another. It is also a hymn to biodiversity'.[28] It's a secular hymn, but nonetheless replete with worship of nature. Beyond the mundane business of telling us how to get birds' names right, field guides induct us into the dazzling diversity and awesome beauty of life on this planet.

By guiding us through some of the myriad mysteries of nature, and making them to some degree comprehensible, field guides allow us to touch a world beyond ourselves. They don't provide answers to any ultimate questions, nor even ask such questions. Yet they do help satisfy a primal human desire for connection with nature. That they do so in distinctively modern ways, entailing comparison, evaluation, and naming, doesn't detract from their ultimate purpose of allowing us to touch the wild.

CHAPTER THIRTEEN

Jacky Winter: Names

THE JACKY WINTER IS a doubly blessed bird. Its vernacular name is among the most amiable on the Australian bird list, and specially suits this charming little melody-maker of the woodlands. *Microeca fascinans*, its scientific name, is exceptionally fitting, too. '*Microeca*' translates as 'little house', a reference to the tiny nest that the members of this genus build; '*fascinans*', from the Latin *fascino*, means to enchant or bewitch. So the two together could be colloquially rendered 'enchanting dweller in a little house'.[1] Few Australian birds are so fortunate as to have both an evocative vernacular and an engaging scientific name. In this chapter, I focus on the vernacular side, especially the quest to find apt and appealing appellations.

That quest goes back a long way. John Leach devoted many words in his 1911 field guide to the topic. 'We need good descriptive names for our varied and beautiful birds', he entreated, 'more children's and poets' names, and less of the deadly formal "Yellow-vented Parrakeet," "Blue-bellied Lorikeet," and "Warty-faced Honeyeater" for some of the most glorious of the world's birds'.[2] Names, to Leach, were far more than mere labels of convenience. Like many birders at the time, he valued attractive names because they fostered feelings for birds, thereby promoting their preservation. Unlike most of his fellow birders, Leach was in a position to do something about this. He convened the Checklist Committee of the RAOU during the decade leading up to its issue of the 1926 *Official Checklist of the Birds of Australia*.

The 1926 RAOU checklist superseded that of 1913, which had given *Microeca fascinans* the drab and dowdy name 'Australian Brown Flycatcher'. It was transmuted into 'Jacky Winter' in the 1926 checklist, which changed some other vernacular names as well. 'Ground-Bird' became 'Quail-Thrush', and some 'long formal names such as Great Brown Kingfisher' were changed to the more familiar 'Laughing Kookaburra'. However, the name changes were neither comprehensive nor systematic, and the 1926 Checklist Committee was well aware that, in words bearing the stamp of its convenor: 'Appropriate, descriptive, children's, and poets' names are still required for many Australian birds.'[3]

Not everyone agreed with that judgement. In a discussion of vernaculars at the 1948 RAOU congress, Charles Bryant 'decried a tendency to introduce childish names'. Eric Sedgwick backed him up, stating that, 'Contrary to oft-expressed opinion, children rarely evolved suitable bird names.' Bryant and Sedgwick, along with several other senior RAOU members, maintained that stability mattered far more than aesthetics in the naming of birds and 'that the general principle should be to alter nothing unless essential'.[4] Cumbersome names, once accepted, should stand.

Bryant seems to have considered enthusiasm for name-changing a species of self-indulgence. 'As I do not set myself up as an ornithological iconoclast', he wrote, 'I am entirely unperturbed by the fact that one of our smallest birds is burdened with a quadruple cognomen, namely Golden-headed Fantail-warbler.' A name is 'merely a label', he insisted, and just as human surnames are not meant to communicate the personalities of their bearers, so bird names need not convey the characters of the birds themselves.[5] It is impossible to ascertain how large a segment of the birding community held to such utilitarian views on names, but it was a substantial segment.

On the other hand, many prominent birders continued to uphold Leach's espousal of the need for attractive children's and poet's names. Alec Chisholm was one, and, like Leach, he held positions pertinent to the matter, being appointed to the RAOU Checklist Committee in 1919 and convening the Vernacular Names Committee from its inception in 1938.[6] He took the latter committee's task very seriously, being committed to finding appealing and appropriate names—such as 'Jacky Winter'—to 'strengthen neighbourly relations between birds and man'.[7] It wasn't only amateur birders such as Chisholm who championed that ambition. In 1977, Dom Serventy told fellow professional Richard Schodde that he concurred with Leach's *Australian Bird Book*, 'wherein he hoped that the common names of Australian birds would emanate not from committees but from school children and poets'.[8]

A year later, a committee chaired by Schodde issued the RAOU's 'Recommended English Names for Australian Birds'. It showed no sign of heeding Serventy's advice, and paid only lip service to Chisholm's Vernacular Names Committee. That committee had been abolished some years earlier, its convenor complaining that the RAOU had disregarded its recommendations.[9]

In its 1978 'Recommended English Names', the responsible committee clearly set out the principles behind its deliberations. Consistency with 'international usage' was prominent, those two words being persistently reiterated throughout the document. Advertising the virtues of its internationalism, the committee maintained that 'an interest in the outside world is a hallmark of a liberal mind and a measure of the cultural level of a people. So it behoves us not to press too far our independent naming of birds that occur in Australia'. It stressed that its recommended names were for '*use in ornithological literature*', and it acknowledged that the public might use other colloquial names in everyday conversation.[10] Nonetheless, the committee's paramount aim was to establish a set of standardised common names that would be adopted by the birding community, professional and amateur, national and international.

Some birders, both professional and amateur, decried that ambition. Dom Serventy told Schodde that imposing a 'universal system of English names' was tantamount to straightjacketing the language, adding that 'we as zoologists have no real business with common, or vernacular names at all'. Scientists should concern themselves with scientific nomenclature, Serventy insisted, and let common names 'evolve naturally'. He also told Schodde that it had recently 'dawned on me why this modern fetish for universal names has come about. It is primarily to cater for the globe-trotting list maker'.[11] This was true insofar as the recent growth of international birding tourism and twitching had energised a push for universal common names. But the move toward standardisation went back much further than that.

Leach himself had urged standardisation of vernaculars, offering his 1908 *Descriptive List of Birds Native to Victoria* not only as an aid to identification, but also 'to help secure uniformity of names'.[12] That aspiration clearly informed the choice of vernacular names in the RAOU's 1926 checklist, which was compiled under Leach's chairmanship; and consistency with international usage was among the principles that guided the Checklist Committee's deliberations.[13] It is equally clear that the committee was guided by Leach's hope that 'appropriate, interesting, inspiring and euphonious names should be used' as vernaculars wherever possible.[14] That aspiration seems to have been pushed into the background by the committee that issued the 1978 'Recommended English Names' while considerations of international consistency were thrust to the fore. The shifts of emphasis testify to both the accelerating pace of globalisation and the growing predominance of professional scientists in the birding community.

The 1978 RAOU English names committee didn't always justify their choice of name with consistency or conviction. Although they replaced 'Heathwren' and 'Warbler' with the generic names 'Hylacola' and 'Gerygone', on the grounds that the former names wrongly implied relatedness to Old World wrens and warblers, only three paragraphs later they endorsed 'Speckled Warbler' as the 'inescapable' name for *Sericornis sagittatus*. A little later on the same page, they gave the tick of approval to 'Inland Thornbill', because it seemed 'the best of a bad lot of possible names'. In a very few cases, the committee accepted names purely on the basis of tradition. Although every other member of its genus was called a flycatcher, *Microeca fascinans* remained 'Jacky Winter' because that name was 'so traditional that it ought not to be put aside'. Although *Rhipidura leucophrys* is not related to the Old World wagtails, and all other members of its genus were called fantails, the committee retained 'Willie Wagtail', and didn't even bother to explain why.[15]

'Jacky Winter' and 'Willie Wagtail' stirred no controversy because these were familiar and well-loved names. New names were a different matter. Many birders were aghast at some of the committee's choices, and none riled them more than 'Thick-knee' in place of 'Stone-curlew'. The committee recommended 'Thick-knee' partly because it was 'coming into fashion' overseas, and partly because 'Stone-curlew' was taxonomically misleading, although they acknowledged that the new name was not anatomically accurate. 'If we cannot have a colloquialism like "wilaroo"', they pleaded, 'an anatomical seems better than a taxonomical misnomer.'[16] It was hardly a ringing endorsement of a name that the committee knew was controversial. In any case, why 'Wileroo' could not have been adopted was not explained.

'Thick-knee' was the most vehemently reviled new name, but 'Baza', 'Gerygone', and 'Calamanthus' weren't far behind. Complaints ranged widely, from the ugliness of the new names to their elitism, from the names' meaninglessness to their robbing the language of diversity and whimsy. One of the more extreme expressions of revulsion came from Cairns BOC member Marion Cassels, who, upon encountering 'Orange-footed Jungle Fowl', 'Wandering Whistling Duck', and 'Pacific Baza', felt she 'could not look any further, or I would be physically ill'.[17] Fellow BOC member Roger Thomas voiced less visceral but still vehement disfavour, contending that standardising common names needlessly duplicated the purpose of scientific ones. If scientists wanted to invent common names, he suggested, 'let them do so in their own journal, leaving the amateurs alone'.[18]

Thomas's words point toward one possible reason for BOC members disavowing the RAOU's recommended English names. The BOC had evolved along different lines to the RAOU, remaining more a birdwatching than an ornithological organisation. In any event, the RAOU recommendations prodded the old sore of amateur–professional tensions. Articles in the BOC's *Bird Observer* magazine in the late 1970s and early 1980s, even those not devoted to nomenclature, often carried asides mocking the new names. An article in the May 1980 issue, for example, revelled in the wonderful birds that members had seen during their tour of the Kimberleys, but embedded a little jibe in the list, recounting how the 'Blue-winged Kookaburras laughed raucously at our clumsy bird nomenclature'.[19]

Yet some BOC stalwarts strongly endorsed the RAOU's recommended English names. Former club president Howard Jarman maintained that the RAOU was right to follow international precedents in naming our birds; to do otherwise would reduce us to 'a bunch of self-satisfied Ockers isolated in our own little corner of the world'. He lauded the 'satisfactory compromise' that the RAOU names committee had achieved in 'bringing uniformity and stability where diversity and confusion currently prevail'. 'Diversity', in the 1970s, had not yet been elevated into a term signalling automatic approbation. Acknowledging that birders had local and individual naming preferences, Jarman insisted that 'prejudices or affection for old, familiar names must be put aside and the recommended names adopted for the sake of uniformity and national usage'.[20] BOC member Alastair Morrison was blunter, condemning criticism of the new names as 'puerile in the extreme'.[21]

While supporters maintained that adopting the recommended names would bring order out of chaos, critics contended that the plethora of unfamiliar and unwieldy titles would worsen the nomenclatural mess and diminish our appreciation of birds. Beyond the disagreements over particular names lay a philosophical difference between those adhering to a utilitarian view of names as mere useful labels for things, and those of a more romantic inclination who saw names as vehicles to help bridge the divide between people and birds, emotionally connecting us to them. Graham Pizzey was of the latter view. He had the misfortune to publish his field guide, representing 15 years of hard work, just after the RAOU issued its recommended names. Some welcomed his failure to toe the line on those recommendations. Allan McEvey, curator of birds at the National Museum of Victoria, congratulated Pizzey for his wonderful field guide, adding that he was 'pleased ... that it has not been blighted by such unwelcome nonsense as Thick-knees, Needle-tails or that wretched super-market Americanism, "chicken" for Domestic Fowl'.[22]

In the introduction to his 1980 guide, Pizzey explained that he did not follow the RAOU recommended names partly because his guide was already with the printer in 1978.[23] That wasn't good enough for many reviewers. Indeed, Pizzey's failure to adopt the RAOU's recommended names was the most common target of complaint in an otherwise overwhelmingly positive reception of his guide. Ornithologist Roger Jaensch, recognising how important field guides could be in swaying the popular acceptance of names, lamented that Pizzey's choices 'will surely erode any success the RAOU list has achieved in standardising English names'. 'Sacrifices of personal preferences are required', Jaensch declared, with a clear implication that standardisation trumped all other considerations in the naming of birds.[24] John Penhallurick slammed Pizzey's disregard of the RAOU's recommendations as an 'unfortunate decision [that] will badly set back the cause of uniform names, to the disadvantage of every birdwatcher'. He also questioned the validity of Pizzey's excuse that the RAOU list appeared after his book was in press: 'since two and a half years have elapsed between the appearance of the names and Pizzey's book, it is hard to believe that he could not have switched to the new names if he had wished to do so'.[25]

Penhallurick had a point. Pizzey didn't use the new names, not only because his book was in press when the names were issued, but also because he detested them. In September 1982, he sent Schodde a list of bird names he would not adopt in future editions of his guide, with clear—and sometimes blunt—explanations why. Against 'Buff-breasted Paradise Kingfisher', he wrote 'Too bloody much', while 'Feral Chicken' was annotated 'Its idiocy is self-evident'.[26] Writing to ornithologist Allen Keast, who shared many of his views on names, Pizzey explained his preferences more temperately. From his experience as a journalist and field guide author, he had come to understand that 'what the lay public want ... are simple, memorable and if possible explanatory names that tell something about the bird in question, even if that information is taxonomically imprecise'. Common and scientific names served 'different, and, in a sense, opposing purposes' he believed, so there was no point in being scientifically pedantic about vernaculars. Reprising the conservationist theme that infused all his work, Pizzey argued that the choice of name was bound up in the bigger project of 'trying to make the Australian public more aware of birds, and through birds, the environment generally'.[27] Names that did not help people connect with birds were wasted opportunities.

Names have been common causes of contention among birders, and the 1978 'Recommended English Names' raised dispute to new heights. So when the RAOU revisited this delicate topic in the mid-1990s, in preparation for the issue of a new checklist, they held a plebiscite on the 24 names in the 1978 list that had provoked the fiercest opposition. Opinions were sought from members, not only of the RAOU, but also of other birding groups, including the BOC. To a far greater extent than for the 1978 list, the RAOU strove to involve the birding community in the decision-making process—or at least to give a semblance of involvement. Throughout 1993 and 1994, the RAOU magazine, *Wingspan*, carried articles explaining the naming process, one of which acknowledged that the committee responsible for the 1978 recommended English names had made mistakes, sometimes tying English and scientific names too tightly together, and being inconsistent in applying the relevant principles.[28]

The plebiscite produced some predictable results. 'Stone-curlew' topped the poll with 1,133 votes, while its formerly imposed name, 'Thick-knee', got only 164. 'Black-fronted Dotterel' trumped 'Black-fronted Plover' almost as comprehensively, with 1,048 for the former against 219 for the latter. Less massively, though still decisively, 'Heathwren' outpolled 'Hylacola'; 'Fieldwren' beat 'Calamanthus'; and 'Nankeen Kestrel' bettered 'Australian Kestrel'. There were some surprises. 'Black-necked Stork' got almost twice as many votes as 'Indo-Pacific Jabiru' (probably the 'Indo-Pacific' bit doomed it), and 'Brown Cuckoo-Dove' polled ahead of 'Brown Pigeon'. In two cases, the voting was unexpectedly close. 'Pacific Baza' got 48.1 per cent of the votes, while 'Pacific Crested Hawk' got 51.9 per cent. 'Gerygone' was endorsed by 46.6 per cent of voters, 'Warbler' by 48.6 per cent, and 'Flyeater' by 4.8 per cent. Partly because these results were so close, the RAOU stuck with 'Baza' and 'Gerygone'. Three other names were left unchanged pending further discussion, but by and large the RAOU adopted the names endorsed by the 1994 plebiscite.[29]

There was some debate over the new choices of name, but nothing like the furore that had erupted in 1978. The tactic of following some semblance of democratic process seems to have worked. Not everyone was entirely happy with the outcome, but the most reviled names of the 1978 recommendations were revoked, and familiar ones reinstated. There were some gentle jibes about the silliness of a naming process that seemed to go around in circles, but very few vehement objections to the names adopted in the mid-1990s. Among the few was a letter by John Squire in the March 1995 issue of *Wingspan*. He attacked not the names themselves, but rather the process of imposing standardised, internationally ratified English-language names that displaced local vernaculars and diminished cultural diversity. It was an old argument, but Squire framed it in the modern idiom of 'cultural imperialism'.[30]

Among those stung to respond to Squire's tirade was Les Christidis, the ornithologist who, with Walter Boles, had published *The Taxonomy and Species of Birds of Australia and its Territories* in 1994. For its English-language names, this work incorporated the outcomes of the RAOU inquiries into birders' preferences, replacing unpopular names such as 'Thick-knee' and 'Pink Cockatoo' with 'Stone-curlew' and 'Major Mitchell's Cockatoo' respectively. Christidis explained the usefulness of standardised names, for example in lobbying public figures for bird conservation, and maintained that their adoption 'will not lead to a loss of culture'. He concluded by suggesting that, 'J. Squire should learn to appreciate the beauty of birds and not get so carried away with what they are called.'[31] Maybe so, but getting 'carried away' with what birds are called seems to be written into the birding script, and shows no sign of going away.

Today, a continuing controversy over bird names concerns those that commemorate individual agents in the colonisation of Australia. The example most often invoked is 'Major Mitchell's Cockatoo', named after a 19th-century explorer who sometimes clashed with Aboriginal people. This name, according to a segment of birding opinion, should be changed back to 'Pink Cockatoo', and all other allegedly culturally inappropriate names abolished.[32] Among the names that would suffer this fate are 'Bourke's Parrot', 'Regent Honeyeater', and 'King Parrot' (named not after the monarch, but after governor Philip Gidley King). Many birders prefer to retain those names, since they carry a patina of familiarity. Regardless of the rights and wrongs of either side, those urging the name changes, in this as in all previous such disputes, clearly concur with the notion that names are not mere labels of convenience, but potent contributors to how we evaluate the world.

CHAPTER FOURTEEN

Straw-necked Ibis: Reputation

I MUST HAVE SEEN them before, but the first time I identified a Straw-necked Ibis was in February 1965. A flock of 20 or so were probing around the vacant lot between Eventide Home for the Aged and the Rockhampton State High School, just a block away from home.

I was ten and a total novice at birdwatching. I'd recently bought a copy of Cayley's *What Bird Is That?*, but didn't have enough pocket money left to buy a pair of binoculars. So instead, I bought a little pocket telescope, and it was through that dinky device that I now watched the ibises. I saw something I'd never noticed before, and needed Cayley's prompting to see. Below their necks they had streamers of straw-coloured plumes. It was not a momentous observation, but it hit me as a revelation.

Over the next several years, I continued my inexpert birdwatching, eventually acquiring a cheap pair of binoculars and supplementing Cayley with Leach's *Australian Bird Book*. With those tools, identifying a bird could still be a challenge, but I eventually learned how to tell a Grey Teal from a Black Duck.

Not all birders began the hobby in their childhood, but many did. Some went on to adult careers in ornithology. Just after his election to the presidency of the RAOU in 2001, Henry Nix recalled his introduction to that organisation as a 13-year-old boy from Ipswich:

> A letter that I had written and that was published in Crosbie Morrison's *Wildlife in Australia* evoked a response from Alec Chisholm. Astonished and indeed flattered by a letter from one of the best-known bird men in Australia at that time, I sought his advice on how I could learn more about birds. He suggested various readings (including, of course, all his own books) and urged me to join the RAOU. Funding the annual subscription in those early years was difficult, but I owe Alec Chisholm a debt of gratitude.[1]

From those boyhood beginnings, Nix went on to become not only RAOU president, but also an internationally renowned ornithologist, specialising in computer-based inventory and evaluation systems for environmental management.

Nix's boyhood inspiration, Alec Chisholm, also began birding as a boy, although his youthful ventures were bloodier than Nix's. Growing up in the Victorian country town of Maryborough around the turn of the 20th century, Chisholm's boyhood birding entailed nest-robbing and killing birds with stones and a shanghai. Even when he turned to bird observation, blood was often shed in the process. He began keeping a nature diary in 1907, at the age of 17, and its early entries record his shooting birds, including Purple-crowned Lorikeets and Spotted Quail-thrush, for identification and for specimens.

Also at the age of 17, Chisholm joined the two leading Australian

birding organisations, the Australasian Ornithologists' Union and
the Bird Observers' Club. In 1908, he attended his first BOC meeting,
and there met journalist Donald Macdonald, whose genial writings
on nature had been a powerful inspiration behind Chisholm's
own development as a naturalist. At that meeting, Chisholm
recollected, 'I began by confessing that my early interest in birds was
accompanied by a shanghai', but Macdonald immediately put him at
ease. '"Don't worry, lad", he laughed, "that's the way we all began."'[2]

Macdonald was right, but at the time he spoke moves were afoot
to ensure that killing would no longer be 'the way we all began'
birding. In 1909, the Gould League of Bird Lovers was founded in
Victoria to inspire in children a love of birds and a commitment to
their conservation. Its members pledged to protect native birds, not
steal their eggs, and to try to prevent others from harming them in
any way. An immediate success, the Gould League soon expanded, to
New South Wales in 1910 and later to all Australian states. Elsewhere,
the leagues' fortunes waxed and waned, but in Victoria and New
South Wales they held strong through most of the 20th century.[3]

Prominent among the league's founders was John Leach. His twin
roles as school inspector and pioneering field guide author epitomise
the partnership of birders and teachers that gave birth to the Gould
Leagues and sustained them thereafter. Leagues operated through
the state school systems, with teachers conducting many of their
day-to-day activities while birders, including members of the RAOU
and the BOC, provided ornithological guidance and expertise. It
was a fruitful collaboration with a grand ambition: to inculcate a
conservation consciousness in the Australian populace by alerting
the rising generation to the wondrousness of the birds around them.

'Education is more potent than legislation in the matter of bird
preservation.'[4] That was the motto of the New South Wales Gould
League, but all state leagues worked on the same assumption. By

'education' the league did not mean the 'chalk and talk' for which schools of the day have since—perhaps unjustly—become infamous. Gould League education was participatory and experiential, giving children contact with wild birds, and inducting them into experiencing nature at first hand. Sometimes, lessons were literally hands-on, with children being encouraged to feed birds from their hands and feel the ticklish touch of claws on skin. There was even a Gould League merit badge awarded to 'any member who can induce a wild bird to alight or perch on his or her person'.[5] The aim, as a 1934 league publication put it, was to bring 'us into more intimate contact with our feathered friends'.[6]

In furtherance of that aim, the league instituted an annual Bird Day, first celebrated in Victorian state schools on 29 October 1909. It spread to other states as the league expanded. In characteristic Gould League fashion, Bird Day celebrations brought birds into the classroom and took the classroom out to the birds. In the classroom, children wrote essays on birds, recited bird poetry, sang songs about birds, and listened to visiting naturalists extolling the wonders of birdlife. Outside the classroom, children fed birds and built feeding-tables to facilitate the process; they planted bird-friendly gardens, and went on birding excursions into nearby bushlands and parks; and they practised bird-call imitations, sometimes in raucous competitions. As one of its Victorian presidents explained, 'members of the Gould League are active participants in their own learning', thereby deepening their appreciation of 'the value of bird life to mankind'.[7]

Walter Finigan, a founder of the New South Wales Gould League, conceived its mission as:

generating, especially in the children, an abiding, purposeful, and fructifying interest in the natural marvels that lie around us; of opening a most interesting field of observation, full of movement, variety, and beauty which will touch the springs of emotion and further develop the moral and aesthetic needs of our nature.[8]

This melding of experiential, empirical, emotional, and aesthetic appreciations of birdlife typifies birding more generally. It evidently appealed to children, because membership of the leagues increased astoundingly. Within two years of its foundation, the New South Wales league had over 17,000 members, and by the 1930s it was enrolling over 100,000 children annually.[9] Numbers were comparable for the Victorian league.

As part of their program of participatory appreciation, Gould Leagues encouraged children to take up birdwatching as a hobby, and gave practical guidance on how to do it. Advice was along conventional natural history lines, stressing the need for close observation, careful identification, and accurate note-taking.[10] However, the empirical side of bird study comprised just one component of a holistic appreciation of birds. The leagues were set up not as training grounds for birdwatchers, still less for ornithologists, but rather to foster a widespread love of birds. Empirical knowledge and scientific understanding were highly valued, but children's emotional, ethical, and aesthetic relationships with birds came first.

The *Bird Lover*, the magazine of the Victorian Gould League, was filled with articles rejoicing in children's friendships with birds, many written by the children themselves. In the 1963 issue, for example, the pupils of Allambee South State School expressed their delight in interacting with birds in their schoolyard, especially having them perch on hands and shoulders, sometimes to be fed, sometimes just to sit there. 'This morning I had a thrush sitting on each hand and I felt very proud', one pupil confided. Another reported that an Olive Whistler 'looked at me and flew straight down quite unafraid and stayed on my hand for five or six minutes'. When they found a fledgling Mudlark that had been blown from its nest, the children named it 'Joka' and raised it to maturity, feeding it on pieces of worm and minced steak. The article was illustrated with three photographs by Miss R. Jones (presumably the teacher): one of a girl feeding Joka, another of Joka standing on a boy's hand, and the third of Joka perched on a girl's shoulder as if whispering into her ear.[11] The clear intention was to encourage intimacy between birds and children.

The league also took children further afield to enjoy contact with birds. Mainly on Bird Day, but sometimes on other days, thousands of pupils from Melbourne and Geelong schools were bussed out to bushy places such as Toolern Vale, Eltham Park, and the You Yangs to give them experiences unavailable in the city. Reports on their visits, by the children themselves, were published in the *Bird Lover*, often with accompanying photographs. An article on the 1959 excursion to Toolern Vale, by grade six students Virginia Jackson and Glenda Brigham, features a charming photo of two other Gould Leaguers, Dawn Fisher and Margaret Moran, creeping through the grass to photograph birds with their Box Brownie.[12] As this and numerous other *Bird Lover* articles attest, girls were prominent participants in all Gould League activities, from writing poetry about birds to observing them in the bush. Gould Leagues deliberately reached across the gender divide.

Love of birds, personal interaction with birds, and field observation of birds all dovetailed into the Gould Leagues' commitment to bird conservation. There was also a more pragmatic side to how that commitment was advanced. Like other conservation bodies established in the early 20th century, Gould Leagues placed considerable weight on the usefulness of the living things they sought to safeguard. As long-term secretary of the Victorian league H.N. Beck put it, the league's purpose was to ensure people would 'fully recognise the economic as well as the aesthetic value of birds, and so would voluntarily protect them'.[13]

For economic value, no species outdid the Straw-necked Ibis, although its White cousin ran a close second. Other birds, such as Australian Magpie and Laughing Kookaburra, were praised for their pest-destroying propensities in the many publications showcasing species that exemplified the league's dictum on the worthiness of birds. But for usefulness, the ibis was the standout.

An article in the 1953–54 *Bird Lover* by schoolboy David Street began: 'The straw-necked ibis has the reputation of being the farmer's best friend. This awkward-looking bird thoroughly deserves this title because it daily disposes of immense quantities of grasshoppers.' Marvelling that a single ibis could consume 3,000 grasshoppers a day, Street added the equally marvellous observation that, 'Most birds that do good often do some harm, but the ibis doesn't seem to do any harm.'[14] A 1959 cartoon sequence by sixth-grader Kerin Swinburne reprised the theme, depicting a flock of ibises descending on a field to devour the pesky grubs. 'Thank you, Mr Ibis', says a farmer while shaking an ibis's foot in gratitude in the final frame.[15]

Prominent birder and vice-president of the Victorian Gould League, Roy Wheeler, tackled the theme more scientifically. In a *Bird Lover* article that ranged across the natural history of the species, he proclaimed the Straw-necked Ibis 'the most useful bird as far as the destruction of agricultural pests is concerned'. The bird 'enjoys protection throughout all Australian states', he acknowledged, but this was 'not enough unless protection of their breeding areas is also assured'. One of the big ibis breeding sites in South Australia, Bool

Lagoon, was then under threat, so Wheeler took the opportunity to explain to his young readers why destroying this habitat would be disastrous for the ibis.[16] This was typical of the genial manner in which the Gould League conveyed its conservationist message to children, moving deftly from positive stories about birds to explaining why they and their habitats had to be protected.

This was relatively easy for the ibis, since its status as the 'farmer's friend' was an established truism. A.J. Campbell's 1908 presidential address to the RAOU canvassed the question of which was 'the most useful bird in Australia'. There were many contenders, he acknowledged, but concluded unequivocally that 'the Straw-necked Ibis is by far our most useful bird, both as to its numbers, the amount of food it devours, and not being antagonistic to human interests'.[17] Leach, in his *Australian Bird Book*, ranked the Straw-necked Ibis 'a valuable asset to Australia' for its 'insatiable appetite for grasshoppers and other insects'.[18] So famed were ibises for their usefulness that when, in 1925, an Australian White Ibis was blown across the Tasman Sea in a gale, New Zealand birder Perrine Moncrieff expressed a hope that more might be blown there and 'the birds given a chance to become permanent residents of New Zealand'.[19] So the encomiums continued. According to an article on ibises in a 1941 issue of *Wild Life*, 'it may be fairly said that there is no other type of bird more useful to man'.[20]

Only a few disagreed. In line with his broader stance on bird conservation, Charles Belcher in 1914 repudiated the idea that their status as the 'farmer's friend' made ibises especially deserving of protection. That, he argued, wrongly prioritised economic over aesthetic worth:

The food of all the Ibis tribe consists largely of grasshoppers and other ground-dwelling insects; they are, of course, not plentiful enough to make any appreciable difference to the insect pests of these parts, and are rather to be preserved as handsome and interesting birds than on account of any special economic value.[21]

Few of Belcher's birding contemporaries would have disputed his depiction of ibises as 'handsome and interesting birds', but most recognised that their economic value provided more solid grounds on which to campaign for their protection.

Gould League writings about ibises acknowledged their aesthetic appeal alongside their utilitarian virtues. In a 1962 item on the Straw-necked Ibis, Colin Grant of Yallock State School described the bird as 'graceful' and 'beautiful', as well as praising it as 'a very

DRAWN BY JOHN MIDDLETON, S.S. 4783, KANANOOK

Figure 18. Back cover of the 1969 issue of *The Bird Lover*.

useful bird because it destroys the caterpillars which are so harmful to pastures and crops'.[22] A charming painting by John Middleton of Kananook State School, published on the back cover of the 1969 *Bird Lover*, showed two White Ibises flying over a rural landscape, under the caption 'Graceful Guardians'.[23] The two words encapsulate the twin qualities of beauty and usefulness that were deployed together by the Gould League for conservationist ends.

As well as celebrating the ibis in prose and pictures, the league also lauded it in verse. C.R. Forster of Sidebottom via Taree won first prize in a 1937 New South Wales Gould League poetry competition for an entry titled 'The Straw-necked Ibis', whose third stanza acclaims the bird's usefulness in conventional imagery:

Content to loiter in sequestered ways,
Nor yet complain 'gainst man's progressive hand;
They garner where the ploughs the furrows raise,
Or glean a thousand pests from pasture land.
Their slender sickle-beak's unceasing toil,
Unbidden serves both husbandman and soil.

However, the other five stanzas never mention usefulness, instead delighting in the ibises' 'graceful flight', their 'happy daily bivouac', and their 'mating call! The strange behest of Spring!'[24] Even for ibises, whose reputed usefulness surpassed that of other birds, Gould League publicity drew attention to the birds' aesthetic qualities and their intrinsic worth as living things. The 'farmer's friend' was also a handsome denizen of our lands.

In recent years, such positive representations have suffered a setback. The ibis image-problem began when the White species intruded into a habitat many humans regard as peculiarly their own. From the 1980s, White Ibis transformed themselves into urban birds, thriving not only in the suburbs, but also in the CBDs of Brisbane, Sydney, and other cities. They previously had visited those places, but now they became residents. Preying on hamburgers instead of grasshoppers, striding across café tables instead of paddocks, rummaging in rubbish bins instead of furrowed fields, the birds became nuisances in the eyes of many. In its new urban home, the White Ibis morphed from pest-destroyer to pest, and earned the derogatory sobriquet 'Bin Chicken'.

But things are never straightforward in human-avian relations. Though denigrated by some, urban ibises retain a popular following. In the 2017 *Guardian Australia's* Bird of the Year competition, the White Ibis came second with 19,083 votes, just behind the Magpie's 19,926.[25] In his investigations of public attitudes among Brisbane residents, urban ecologist Darryl Jones found not only strong community support for the birds, but also an exceptionally enthusiastic coterie of ibis devotees who call themselves 'The Disciples of Thoth' (after an ibis-headed god of ancient Egypt).[26] There's probably a tongue-in-cheek element in that name; but there is, too, in many of the popular pejoratives for the White Ibis: 'Tip Turkey', 'Picnic Pirate', 'Refuse Raptor', and even 'Bin Chicken' itself. They're derogatory, but with an edge of humour that betrays affection.

Indeed, affection is at least as strong as disdain. When the White Ibis transformed itself into an urban bird it entered modern, urban popular culture, and quickly attracted the diversity of feelings and attitudes typical of popular culture icons. Some loved it; some loathed it. Some had ibises tattooed on their torsos; others participated in International Glare at Ibises Days. Many ibis cultural artefacts are tawdry, even tacky, but their proliferation attests to the extent to which the White Ibis commanded the attention of urban Australians.[27] Nonetheless, from 'farmer's friend' to 'Bin Chicken' is a big demotion.

Recently, the White Ibis may have found a way to redeem itself. It's by being useful in helping to control a devastating pest species—the introduced, poisonous cane toad. In 2017, a Brisbane birder noticed White Ibises catching cane toads and alternately bashing them and washing them, then swallowing them whole.[28] The birds had learned how to provoke the toads into releasing the toxin from their neck-glands, then cleaning it off before eating them.

When their toad-eating skills were widely publicised in late 2022, one news outlet reported the fact under the headline 'This Just In ... Ibis have a Purpose in Society!'[29] That's journalistic sensationalism, but other media outlets also homed in on the White Ibis's usefulness in eating cane toads. It even made international news, with the BBC quoting Sydney-based ecologist Rick Shine on what 'a wonderful job' ibises were doing as part of 'an unseen army that are reducing the numbers of cane toads every year ... So we really should be grateful for some of these unloved Australians'.[30] Echoes of the old 'farmer's friend' image aren't difficult to discern.

Straw-necked Ibises have also been observed using the 'stress-and-wash' method to cleanse cane toads before consumption. Perhaps that will boost their image, too. It's not that Straw-necks are belittled as Bin Chickens; they've remained the country cousins who neither reside in cities nor rummage in rubbish bins. Their problem is that they've largely fallen from public notice, and, unlike the White, have no significant place in the modern Australian imaginary. A role as cane-toad killers may be just what they need to put them back into popular culture. The old Gould League publicity for the 'farmer's friend' may look quaint to modern eyes, but there's shrewdness behind the underlying assumption that usefulness can be a great way to enhance a bird's reputation.

CHAPTER FIFTEEN

Blue-faced Parrot-Finch: Twitching

I'VE NEVER SEEN A Blue-faced Parrot-Finch, though not for want of trying. I've searched the recommended spots in the Julatten–Mt Lewis area four times, and the region around Topaz on the Atherton Tablelands twice. Every time, other birders have unhelpfully told me 'they were here earlier this morning/yesterday/last week'.

Unlike me, Sean Dooley has seen Blue-faced Parrot-Finches. This is because he's a far more skilled and determined birder than I am. As a birder, I'd rank myself in the keen and competent category. Dooley is among the elite of Australian birders: not keen, but fanatical; not competent, but expert. He's also among the most famous of Australia's birders, with a public profile regularly reinforced in newspapers and magazines, podcasts, and tweets.

Dooley first won public acclaim for his 2005 book, *The Big Twitch*, recounting his quest to break the Australian twitching record by seeing 700 species in a single year. Deftly combining adventure with introspection, and humour with insight, his book boosted not only Dooley's own reputation, but also the profile of birding in Australia.

Not least, it trained the spotlight on the variant of birding known as twitching.

Dooley began his Big Twitch the moment the clock ticked past midnight on 1 January 2002. He saw species number 700 at Mt Lewis on 24 December that year. It was a Blue-faced Parrot-Finch. As befits a twitcher, he didn't stop there, but kept seeking rare and obscure species until the calendar clicked over to 2003, before then adding Red-rumped Swallow, Masked Owl, and Little Bittern to his list. His final total was 703. But it was the Blue-faced Parrot-Finch that clinched Dooley's goal of 700: an extraordinarily apt species with which to crown a twitching triumph.

The name 'Blue-faced Parrot-Finch' is quite a mouthful, but the bird behind it is exquisite. Its overall plumage is a lovely grass-green, offset with a cobalt-blue face and red tail. Its colours are bright, but when feeding in its usual habitat of grassy patches on rainforest edges, the bird can be very difficult to see. (So I'm told by those who have seen one.) Add to that the facts that in Australia the species is restricted to a few tiny patches of Queensland's Wet Tropics, and that within those patches it's rare, and the finch's prominence on twitchers' must-see lists is understandable.

So rare is the Blue-faced Parrot-Finch that its presence in Australia was uncertain for many years. The species was not listed in the RAOU's 1913 *Official Check-List of the Birds of Australia*. In a 1925 issue of the *Emu*, G.A. Heuman flatly stated that no species of parrot-finch lived in Australia, all members of the relevant genus (*Erythrura*) being restricted to New Guinea and the islands of the South Pacific. He acknowledged that two specimens of the blue-faced species 'have been obtained in North Queensland, but these were probably casual visitors or escaped birds'.[1] The species was included in the RAOU's 1926 *Official Checklist of the Birds of Australia*; but, six years later, in *Australian Finches in Bush and Aviary*, Neville

Cayley acknowledged that there were only four records of the finch in Australia.[2] When German ornithologist Klaus Immelmann issued an update of Cayley's book in 1965, he could still state that 'there are fewer than a dozen records for the Australian mainland'. Immelmann devoted a year to investigating Australian finches in the wild, during which the only species that eluded him was the Blue-faced Parrot-Finch.[3] This is the kind of bird that twitchers love.

But what's a 'twitcher'? It's a slippery word, but its history might help clarify its meaning. 'Twitcher' and 'twitching' entered birding jargon in the mid-1950s in England, originating in stories about how a young Howard Medhurst would arrive at his birding destinations after riding pillion on a motorcycle, trembling from a combination of cold, craving for a cigarette, and excitement at the prospect of seeing new birds.[4] It's the last of those that endures as the essence of twitching. As British birder and comedian Bill Oddie explains, a twitcher is the kind of birder who, when faced with the possibility of ticking off a new bird, 'is so wracked with nervous anticipation (that he might see it) or trepidation (that he might miss it) that he literally twitches with the excitement of it all'.[5] Oddie exaggerates for comic effect, but 'twitching' has come to denote an energetically competitive form of birding that focusses on adding new birds to a personal list, usually by haring around the country after rarities and vagrants.

Although 'twitching' became birding jargon in the mid-1950s, something very like it had been practised long before. Frank Chapman launched the Audubon Society's first Christmas Bird Count in 1900, and within two decades birdwatching as a competitive sport was well established in the US, led by the doyen of rapid-fire field identification, Ludlow Griscom. In the UK before World War II, there were already 'tally hunters' who did much the same thing as those who were later called 'twitchers'.[6] Australians

William Cooper's painting of Scarlet-chested Parrots captures the dramatic beauty of the birds in their arid habitat. *Courtesy of Wendy Cooper.*

LEFT: The Capricorn Yellow Chat with which this book begins, photographed by Vince Lee on the Port Alma Road, 21 February 2020. *Courtesy of Vince Lee.*

LEFT: Roger Tory Peterson with Rainbow Lorikeets at Currumbin Bird Sanctuary on Queensland's Gold Coast, 1971. Peterson considered Currumbin 'one of the great bird spectacles of the world'. *Courtesy of the Roger Tory Peterson Institute.*

Peter Slater's painting conveys the often-underappreciated gorgeousness of Galahs.
Courtesy of Raoul Slater and Sally Elmer.

Ebenezer Gostelow's paintings of Superb Lyrebirds (above, 1939) and Little and Plumed Egrets (below, 1938) express his enchantment by birds. *National Library of Australia.*

ABOVE: Raoul Slater's photograph of a Plumed Egret celebrates the aesthetic glory of a bird in nature — in this case, a lake on the outskirts of Gympie. *Courtesy of Raoul Slater.*

BELOW: Graham Pizzey's photograph of a Musk Lorikeet in a red-flowering gum pays homage to avian beauty in a different key. *Courtesy of Sarah Pizzey.*

2 Mallee-Fowl	6 King Quail	14 Diamond Dove
3 Stubble Quail	8 Painted Quail	16 Bronzewing Pigeon
4 Brown Quail	11 Plain Wanderer	

384 Spangled Drongo	386 Spotted Bower-Bird	382 Pied Bell-Magpie
385 Satin Bower-Bird	390 Apostle-Bird	394 Gray Bell-Magpie
385A " " (Female)	391 White-winged Chough	

ABOVE: Some of the rather clunky illustrations in Australia's first field guide, John Leach's *An Australian Bird Book* of 1911. *National Library of Australia.*

Plate XIX

BIRDS OF THE FOREST BORDERS AND GRASS-LANDS

LEFT: Neville W. Cayley's illustrations in *What Bird Is That?* (1931) are more artistically accomplished than Leach's, but his plates are crowded and each image is tiny. *National Library of Australia.*

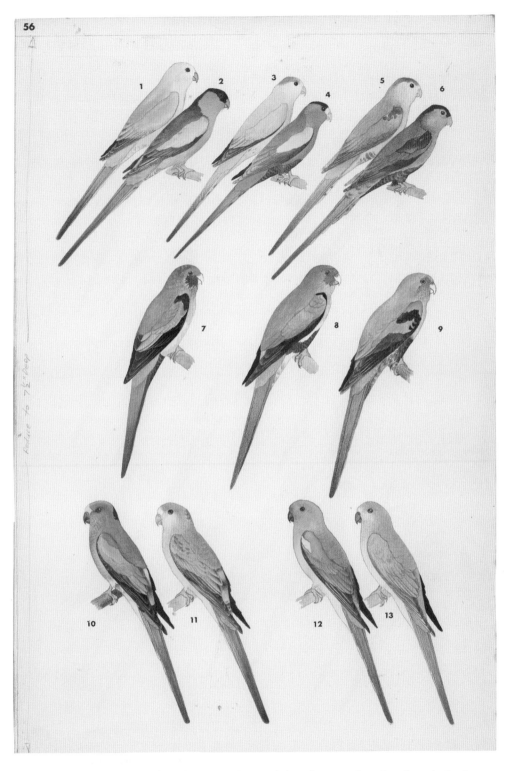

In his ground-breaking *Field Guide to Australian Birds* (1970), Peter Slater lined up the birds, such as these *Psephotus* parrots, in standardised poses to highlight their field marks. *Courtesy of Raoul Slater and Sally Elmer. State Library of New South Wales, PXD 1506, box 8, no.6.*

350
RED-RUMPED
PARROT

351
MULGA
PARROT

354
PARADISE
PARROT

355
BLUE BONNET

race
haematorrhous

race
narethae

352
GOLDEN-
SHOULDERED
PARROT

353
HOODED
PARROT

334
COCKATIEL

150 mm

Graham Pizzey and Roy Doyle in their *Field Guide to the Birds of Australia* (1980) went a step further and included Peterson-style pointers to the field marks. *Courtesy of HarperCollins and National Library of Australia.*

FLAME ROBIN

imm.

♀

♂

SCARLET ROBIN

♂

imm.

♀

RED-CAPPED ROBIN

♂

♀

imm.

In *The Graham Pizzey and Frank Knight Field Guide to the Birds of Australia,* the artist dispensed with pointers but added decorative and diagnostic bits of habitat, as in this plate of red robins. *Courtesy of Frank Knight and National Library of Australia.*

PLATE 113

Plate 125
Page 247

Peter Slater's paintings for the *1986 Slater Field Guide to Australian Birds* splendidly exhibit the combination of art and science that typify field guides. *Courtesy of Peter Slater and Sally Elmer. State Library of New South Wales, PXD 1506, box 18, no.9.*

PLATE 114

Plate 126
Page 249

LEFT: Ebenezer Gostelow painted this portrait of a pair of Paradise Parrots in 1936, which was probably around the time the species fell to extinction. *National Library of Australia.*

RIGHT: William Cooper's painting of Paradise Parrots is typically attentive to the species' habitat. *Courtesy of Wendy Cooper.*

Birds of Australia : Passeriformes ... Atrichornithidae

The Noisy Scrub Bird [Atrichornis clamosus] ♂ 0⅞₇₆₂ cm ・HABITAT : S.W.Aus. only. (Probably extinct)

ABOVE: The lower-right caption on Ebenezer Gostelow's 1933 painting of a Noisy Scrub-bird records the species as 'Probably extinct'. *National Library of Australia*.

RIGHT: In 1961 the Noisy Scrub-bird was rediscovered by Harley Webster, pictured here on left, with Robert Stranger (holding a Noisy Scrub-bird) and Dom Serventy on the right. *Photo by Rica and Sydney Erickson; courtesy of Allan Burbidge.*

THE SPIRIT OF SYDNEY

SCARLET HONEYEATER AT NEST IN SUBURBAN GARDEN

LEFT: Neville W. Cayley's 'The Spirit of Sydney' (1932) beautifully illustrates the accessibility of nature to suburban Australians. *National Library of Australia.*

BELOW: Peter Trusler's 1979 painting of a Common Myna is a cheekily affectionate tribute to another avian denizen of suburbia. *Courtesy of Peter Trusler.*

ABOVE: Raoul Slater's photograph of a common bird – the much-maligned White Ibis or 'Bin Chicken' — reveals its extraordinary beauty. *Courtesy of Raoul Slater.*

ABOVE LEFT: The stunning realism of Peter Trusler's paintings of common garden birds, like this Willie Wagtail, captures their endearing charm. *Courtesy of Peter Trusler.*

ABOVE RIGHT: The Superb Fairy-wren with which this book ends, photographed by me in Turner, Canberra, 27 February 2022.

William Cooper's painting of pastel-plumaged Bourke's Parrots conveys the subtle beauty of the birds, their habitat, and the harmony between them. *Courtesy of Wendy Cooper.*

remained more faithful to a natural history tradition of birding. As late as 1986, Gillian Lord observed that twitching 'is very popular in Britain, and America, but does not seem to have caught on much in Australia'.[7] But she was writing at the very time that twitching was catching on in Australia.

'In some countries "listing" is a recognized sport', Rosemary Balmford explained in her 1980 primer, *Learning About Australian Birds*. Not here, however:

> We in Australia do not seek out places where we will see a lot of birds for the sake of seeing a lot of birds. The rarity-hunting of the northern hemisphere is far less evident here. We go to a particular place to see the birds that live there, and if we see a new bird or an uncommon bird that gives us pleasure; but the object of the exercise is not, normally, only the new or the uncommon bird. The emphasis here is different.[8]

Things had changed by the time Balmford issued a second edition of her book in 1990. 'The sport of twitching has now reached this country', she declared. 'Of course there have always been lists, and most of us keep a list of some kind', she immediately added, but 'twitching as a recognised sport, with twitchathons, and telephone hotlines, is new to us.'[9]

To the innovations listed by Balmford, we could add several more that signalled the advent of twitching in Australia in the 1980s. *Cosmic Flashes*, Australia's first twitching newsletter, began publication in 1981. Around the same time, its editor, John McKean, founded the 600 Club, whose sole qualification for membership was to have seen 600 or more Australian bird species. In the 1980s, for the first time, some Australian birders self-consciously identified as 'twitchers', distinguishing themselves from mere birdwatchers (or 'dudes'). Twitching argot came to Australia in that decade, allowing members of the tribe to identify each other by sharing stories of 'gripping off', 'cripplers', and 'dipping out', perhaps spiced with a little 'stringing'.

There were some tensions between self-proclaimed twitchers and conventional birdwatchers. Peter Slater in 1988 derided twitching as a 'solipsist' form of birding, though he conceded that it might hone observation skills.[10] A few years later, Graham Pizzey belittled the practice, explaining that he had:

> never really been able to appreciate the lemming-like thunder
> of twitcher feet from one end of the country to another to see a
> small bird often perfectly common and everyday in some other
> part of the world, but which by getting caught up in a cyclone
> or a flock of the wrong species, or by boarding the wrong ship,
> has turned up in the wrong place, out of context, often out of
> its true habitat.[11]

Twitcher Andrew Stafford returned the favour, remarking in a review of the Pizzey and Knight field guide that Australian birding was riven by a divide between 'the twitchers and the tea-set'.[12]

Stafford seems to have enjoyed provoking the 'tea-set'. He began one item in *Australian Birding* magazine (the successor to *Cosmic Flashes*) by flaunting the solipsism decried by Slater and deliberately parading the pointlessness of what he did: 'The following account contains no Great Revelations to Science', he wrote. 'Actually, I doubt that it has any redeeming features at all.' His trip to see all of Australia's grasswrens had merely been a holiday that had offered him the opportunity to brag about what he had seen 'for the purpose of maximum self-gratification and ego-massaging'. Thumbing his nose at the birdwatching establishment, Stafford proclaimed himself one of 'the petty pioneers of a new form of utter birding nonsense'.[13] Flippancy and self-deprecation were badges that many twitchers wore with pride.

While there was some sniping between twitching enthusiasts and conventional birdwatchers, twitching was generally accepted into mainstream birding in Australia without too much fuss. There was nothing like the heightened animosity in the UK, where, in the early 1980s, the RSPB refused to have its magazine sullied by 'twitchers', even blacking out the word in an advertisement for Richard Millington's *A Twitcher's Diary*.[14] Australian birding associations were more accommodating. According to Rosemary Balmford, twitching was 'actively promoted by the RAOU'.[15] In 1982, the *RAOU Newsletter* began a Twitcher's Corner, which continued for almost four decades, surviving the newsletter's transformations into *Wingspan*, and then *Australian Birdlife*. By the time it was dropped in 2020, twitching had been so normalised that there was no need to allocate it a special corner.

Perhaps the biggest boost to the acceptance of twitching came through Twitchathons, launched by the RAOU in 1984. Some members were dismayed by their competitiveness, but most were mollified by the fact that Twitchathons raised funds for bird conservation. 'Twitching for conservation' became the standard, even hackneyed, headline for articles on Twitchathons in *Wingspan*. For birders, supporting conservation while having fun is an irresistible combination. Looking back in 2008 on 25 years of Twitchathons, Sean Dooley revelled in 'the gloriously ridiculous nature of a birdwatching race', carefully adding that 'the Twitchathon is first and foremost a fundraising exercise for bird conservation'.[16]

Twitchathons had precursors. In 1967, Roy Wheeler initiated the Melbourne Bird Count Challenge, whose objective was 'to break the

Audubon Christmas Bird Count of USA record of 204 species listed in one day'. The Melbourne teams succeeded, logging 209 species on the inaugural challenge day.[17] Wheeler's proto-twitching activities go back further. In 1946, he titled an article on his successful quest to see 100 species in a day 'Scoring a Century', which breathlessly tallied sightings while making explicit analogies with sport (cricket, in this instance).[18] Reinforcing the parallels with twitching, Wheeler candidly admitted to being a compulsive lister, some of his lists being decidedly idiosyncratic, including one of 'birds seen at a test match at Melbourne'.[19] He may have been unusual for the diligence of his listing, but he wasn't alone. Billie Gill and Fred Smith also engaged in competitive tallying in the 1960s.[20] Doubtless, others did, too.

Striving for ever expanding tallies was certainly not unknown to Australian birdwatchers before the advent of twitching in the 1980s. In 1923, W.B. Alexander, a birdwatcher and entomologist employed in researching prickly-pear control in central Queensland, told Alec Chisholm about the birds he saw around his home at Westwood, adding that, 'My list for Australia now stands at 384 species and for the world at 953 species. I shall have hard work getting another 47. I think though a trip to Cairns might do it.'[21] Clearly, Alexander not only kept both Australian and world lists, but also imagined how he might boost them to numerically appealing totals—1,000 in the case of his world list. It's a low number by today's standards, but his aspiration mirrors that of modern-day twitchers.

It's hard to find any element of twitching that lacks historical precedent. Indeed, the more I looked at twitching through the lens of history, the blurrier became any distinction between it and other forms of birding. Even the archetypal attributes of twitching turn out to be less novel than might be expected. In a 1933 article on 'Seeing "new" birds', Alec Chisholm—who was very much a birder in the natural history tradition—admitted he was 'beset by envy upon hearing or seeing manifestations of joy by enthusiasts who are meeting particular birds for the first time'. The diction is different, but the sentiments are those that Bill Oddie lampoons. Chisholm went on to recount the 1924 RAOU campout at Byfield, where at every sighting of a new bird, 'each man thrilled as a child might thrill if characters in his favourite story-book came to life'.[22] That sounds pretty close to twitching with excitement at seeing a new bird.

Yet Balmford was right to identify the 1980s as the time of twitching's arrival in Australia. An astute observer, she realised that the new-fangled sport shared many commonalities with the established hobby, and that the differences were of emphasis rather than of kind. Twitching's novelty lay in its sharpened competitive edge and the relentlessness of its striving for bigger tallies of ticks. These marked an intensification of aspects of birdwatching that had been present from the pastime's beginning, rather than a radical reorientation of bird-based recreation (although some of its early enthusiasts liked to portray it in the latter mode). The selective intensification of the competitive and aspirational elements was nonetheless significant in shaping the course of birdwatching more broadly.

Modern communications—first the telephone and pager, and later the Internet and social media—fed the competitiveness of twitching. So did modern means of transportation. If twitching was born on a motorcycle, in its mature form it relies on jet aircraft to carry its devotees to the far corners of the planet to add ever more esoteric birds to their lists. The kilometres travelled by dedicated twitchers are truly prodigious and are usually reported boastfully, rather than with unease at the fact that this nature-based recreation relies on petrol-guzzling, pollution-spewing modes of transport.

Indeed, it could be a matter for jokes. The 1997 North Queensland Twitchathon team called themselves the 'Giant Petrol Bills'—which they must have incurred since they achieved an astonishing tally of 243 species in 24 hours.[23] Admittedly, Twitchathons are occasions for fun, and teams typically take humorous names. Nonetheless, there was—and often still is—a remarkably cavalier attitude toward the environmental detriments occasioned by the industrialised pursuit of birds.

Although having fun is the point of twitching, its exponents could take it very seriously indeed. Competitiveness breeds rules, and agonising over the rules of the game has consumed an inordinate amount of twitchers' time. The most fundamental rules—over what constitutes a species—are taken from science, although the vagaries of avian taxonomy allow some wiggle room there. More open to dispute are questions such as which birds can legitimately be included on a list and where the boundaries of regions should be drawn for listing purposes.

On the latter topic, the first issue of *Australian Birding* carried an item headed 'Whither Australia?' It had nothing to do with the social or political direction the country was taking, but rather offered guidance on 'what birds can be added to your list'. Its proposed boundaries for 'Australia', 'Australia and territories', 'Australasia', and 'Australian states and territories' prompted disagreement, with some prominent twitchers proffering alternative boundaries. From there, discussion proceeded to determinations of a bird's 'tickability', dividing the Australian avifauna into A, B, C, and D categories according to how often they were seen, whether they were introduced or native, plus several other criteria. These became so specific that a reference to birds that 'have only ever been found dead above the tideline' carried a footnote: 'Although the British and Irish say *on the tideline*, this seems unnecessarily restrictive and should be extended to include any habitat between the tideline and the tip of Mt Kosciusko.'[24] It's arcane stuff, but competitiveness inevitably spawns quibbling.

Yet the heightened competitiveness of twitching should not be allowed to obscure its continuities with conventional birdwatching or its practitioners' more substantive interest in the creatures they count. John McKean was revered (by twitchers) as 'the father of birding and twitching in Australia' and as a birder who 'championed the hunt for ticks'.[25] He was also a committed conservationist and professional ornithologist employed by the CSIRO and the Northern Territory Conservation Committee. Roy Wheeler was a keenly competitive lister, but also an expert amateur ornithologist who encouraged other birders to develop the scientific side of their pastime. President of the BOC from 1950 to 1953, and of the RAOU in 1964 and 1965, Wheeler's 'greatest achievement was to change the focus of many observers from individual birding to more purposeful field ornithology, emphasising data collection'.[26] Their examples could be multiplied many times over, exemplifying the fact that twitchers, usually, also engage with birds in myriad other ways.

As Mark Cocker has argued, it is misleading to regard the twitcher as a distinct subspecies of birdwatcher. Rather, twitching is an activity in which most birders partake, with varying degrees of intensity and frequency, as just one component of their pastime.[27] Twitching—or something very like it—has been a component of birdwatching all along, although in Australia over the past four decades or so it has grown into a much bigger component than hitherto. But most who twitch spend most of their birding hours doing the things all birders do: watching everyday birds in their local patch, contributing data to citizen science projects, lobbying for bird conservation, and having fun by seeing which birds turn up where. Twitching doesn't preclude everyday birding, and never has.

In her light-hearted account of her adventures 'in attempting to qualify for membership of the 600 Club', Sue Taylor declared herself 'unashamedly a twitcher. Literally. Every time I see a species of Australian bird I have not seen before, I twitch with excitement'. She immediately went on to deny that this meant she was interested in nothing but ticking off rare birds: 'To me, twitching means more than just adding another tick to my lifelist, another feather in my cap, so to speak. It's a joy to know that the birds are still there. For me, every new sighting is a symbol of hope for the future.'[28] Taylor recounted her twitching flippantly, but the flippancy is mere froth, as it is in most twitching literature. Beneath the banter lies an endeavour with which all birders identify, whether or not they wear the twitching label. It's the aspiration to connect with nature.

Some twitchers become so obsessed with ticking off new species that they drain their hobby of nature appreciation—perhaps even of fun. For such fanatical twitchers, immediately a new species has been ticked it's time to find another, paying no heed to what's just been seen. That's the sum of their engagements with birds. I've heard of such twitchers, and read about them, but I've never met one. Just as I've never met a Blue-faced Parrot-Finch. Each must be as rare as the other.

CHAPTER SIXTEEN

Eungella Honeyeater:
Classification and Commemoration

WHILE VISITING THE RAINFORESTS on the Eungella Plateau west
of Mackay in September 1959, J.S. Robertson and F.M. Hamilton
were startled 'by a loud, imperious but quite unfamiliar call'. They
soon saw the bird, and, using the few reference books to hand, they
identified it as 'probably the Bridled or Mountain Honeyeater'.
However, one of those books, *What Bird Is That?*, specified this
species' range as 'Cooktown to Cardwell', so they 'felt that the matter
needed critical checking by further field observation'. The next day
offered just such an opportunity. Finding one of these honeyeaters
on a nest, they captured it and took a photograph, confirming their
identification of the species as Bridled Honeyeater. Following usage
in the 1958 edition of *What Bird Is That?*, they preferred the name
'Mountain Honeyeater'.[1]

A few years later, one of the revisers of that edition of *What Bird
Is That?*, Alec Chisholm, visited Eungella, and reported seeing the

Mountain (Bridled) Honeyeater, *Meliphaga frenatus*.[2] Before he went there, he had remarked that 'the highlands west of Mackay ... are a most promising area for ornithological investigation, especially as former birdmen of the district appear to have confined their activities to coastal areas'.[3] Chisholm was thinking of Eungella as 'a meeting place for northern and southern species', but his remark proved more prescient than he could have imagined at the time. It also helps explain why Eungella held its avian secret so long.

During a 1975 faunal survey of the Eungella region by a combined Australian and Queensland Museums team, a new appointee to the former institution, Walter Boles, mist-netted a pair of honeyeaters. He identified them as Bridled Honeyeaters, a species he had never encountered before. Another member of the museums team collected a specimen of this bird, made the same identification, and stashed it in the museum collection under the label 'Bridled Honeyeater, *Meliphaga frenatus*'. A little later, a young volunteer at the Australian Museum, Wayne Longmore, noticed that this bird differed significantly from other Bridled Honeyeaters in the collection. Indeed, it didn't match any bird in any reference work.[4] Suspicions aroused, in 1978 and 1980, Longmore and Boles conducted further fieldwork in the Eungella area, and collected several more specimens.

It soon dawned on them that this was not a Bridled Honeyeater, but an entirely new species. Their formal description of the new species was not published until 1983, but the ornithologists were confident of their discovery long before then. A specimen collected at Massey Creek on 7 December 1978 became the holotype of *Meliphaga hindwoodi*, the Eungella Honeyeater.[5]

In hindsight, it seems remarkable that this species remained unknown to science for so long. Admittedly, its distribution is very limited. Indeed, the Eungella Honeyeater has the most restricted distribution of all Australia's breeding passerines, being found only in rainforest and wet sclerophyll on the eastern side of the Clarke and Sarina Ranges (which subsume the Eungella Plateau) in east-central Queensland. This is not remote wilderness. It's less than 80 kilometres from a major city, Mackay; it was colonised by Europeans in the 1870s; and since the early 20th century the region has supported a flourishing dairy industry. I presume that the dairy farmers and other locals saw the birds but thought nothing of it, so failed to pass on their observations to ornithologically interested persons. The Birri and Wiri owners of these lands would also have known of the bird, but they, too, failed to communicate their knowledge to science.

Making its long concealment additionally difficult to understand, the Eungella Honeyeater is not exceptionally challenging to find. True, it's not particularly easy to find either. It's a smallish, predominantly grey-brown bird that forages mainly in the upper foliage of tall forests, although it often comes down into the lower storeys. Whether high or low, in rainforest or sclerophyll, its loud, spirited whistles and chortles usually give away its presence long before the bird is seen. In my own many visits to Eungella, I've always heard the eponymous honeyeater before seeing it, although admittedly I've often neither heard nor seen it. Still, it's reasonably common within its restricted range, and the challenge of finding it is not overwhelming.

Up to a point, it's understandable that observers in the 1950s, 1960s, and 1970s confused the Eungella Honeyeater with the Bridled. The two are similar, but far from identical. Most noticeably, the Bridled Honeyeater has a distinctly bicoloured bill, bright yellow at the base and black at the tip, whereas the Eungella species has a uniformly black bill. The Bridled is more robust and has an appreciably different call from the Eungella Honeyeater. The initial failure to distinguish the species can be attributed partly to the shortcomings of the field guides of the day. Leach's and Cayley's were the only ones, and neither provided identification cues with sufficient specificity to prompt a birder into making the kinds of discrimination necessary to distinguish these species. Robertson and Hamilton's account of their 1959 encounter hints that they held suspicions about the bird they saw; it didn't fit neatly with what was in Cayley's guide, but Bridled Honeyeater offered the best fit, so they opted for that. In any event, it's a telling reflection on the state of birding in Australia as late as the 1970s that it took a sharp-eyed museum worker with specimens in his hands to realise that previous identifications were mistaken and that this bird, a reasonably numerous inhabitant of a region close to a major city, was a new species.

Longmore and Boles named the species *hindwoodi* in commemoration of Keith Hindwood, who had died some years earlier. Hindwood never saw the bird named after him, and never visited its Eungella home. His connection was indirect. As honorary ornithologist and research associate at the Australian Museum in Sydney, Hindwood had mentored Wayne Longmore, who honoured his memory in the new honeyeater's name. It's fitting that the last Australian avian species to be scientifically described in the 20th century took a name commemorating Australia's most accomplished amateur ornithologist.

Longmore wasn't the only birder whom Hindwood mentored. Earlier, in the 1930s, he had been among the naturalists who took the young Jock Marshall on birding excursions, honed his identification skills, and taught him the bushcraft that was then a normal part of a birder's training. Other mentors on these excursions, when Marshall was aged in his late teens and early twenties, included Alec Chisholm and Norman Chaffer.[6] Marshall went on to become one of Australia's most eminent zoologists and most outspoken conservationists. His youthful nurturing as a naturalist by a group of older enthusiasts seems to have been a common mode of induction into birding at the time. For that generation, before the advent of sophisticated field guides and high-tech paraphernalia, birding was learned as a craft through hands-on experience guided by expert amateurs. This was generally the case even for those who, like Marshall, later pursued careers as professional scientists.

Hindwood inspired the scientific title of the new *Meliphaga* species, but its vernacular name had very different origins. When Longmore and Boles 'proposed that the English name be Eungella Honeyeater', they explained that its first part 'is an Aboriginal word meaning "mountain of the mists"'.[7] Some have offered alternative translations, such as 'land of clouds'.[8] Whatever the translation, 'Eungella' commemorates the Indigenous custodians of the bird's habitat, the Birri and Wiri peoples. For the honeyeater, the commemoration is somewhat indirect, via the use of an Aboriginal word as a placename. But it's commemoration nonetheless. A few other Australian birds have composite names with one part from an Indigenous language and the other from English: Kalkadoon Grasswren and Pied Currawong, for example. There are also a few stand-alone bird names deriving from Indigenous languages, such as Brolga, Budgerigar, and Galah. But most official Australian bird names are not from Indigenous languages and have no Indigenous component.

Vernacular Australian bird names derived from Indigenous languages are few, but scientific names with that derivation are far fewer. As far as I know, there's only one: *Falco berigora* (Brown Falcon), whose specific name comes from an Aboriginal language of New South Wales, though which language remains unclear.[9] The process of allocating and amending scientific names is governed by a corpus of intricate and inflexible rules, overseen by the International Union of Biological Sciences. While it's possible for new scientific names to incorporate elements from Indigenous languages (provided they are appropriately Latinised), an existing name cannot be changed merely for the purpose of indigenising it. That would violate the rigid rules of scientific nomenclature. Vernacular names, by contrast, are far less constrained by rules and more amenable to indigenisation.

Suggestions to do so go back more than 100 years. In a 1920 *Emu* article, Edward Sorenson, a birder and writer in the mould of Henry Lawson's bush tradition, urged the adoption of Aboriginal bird names. They would promote a love of birds among the wider public, he argued, since they 'are euphonious; they have the virtue of originality, and are much better for general use than many of the popular names now on the bird list'. As a starter, Sorenson offered a list of 53 bird names from (unspecified) Aboriginal languages that could be widely adopted.[10] Other birders—Arthur Mattingley, Les Chandler, and Ted Schurmann among them—made similar suggestions over subsequent decades. However, they were mere suggestions, and there was never sustained pressure for the adoption of Aboriginal bird names at any time in the 20th century. None of the authoritative checklists and similar publications, from the RAOU's first in 1913 through to Christidis and Boles' *Systematics and Taxonomy of Australian Birds* in 2008, paid serious attention to Aboriginal names. Some tried to remedy that neglect, among them ornithologist Ian Abbott, who in 2009 urged the adoption of Aboriginal bird names as 'a modest contribution to racial reconciliation'.[11] Such pleas got a sympathetic hearing, but the adoption of Indigenous names proceeded no further.

One difficulty in adopting Indigenous bird names stems from the linguistic diversity of Aboriginal Australia. At the time of colonisation, there were around 250 distinct languages, each with several dialects; and each language, and often each dialect, had its own names for the birds in its territory. Thus, for example, the Sulphur-crested Cockatoo is 'muurrayn' in the Wiradjuri language of central New South Wales, and 'kiiku' in the Paakantyi language a little further to the west; Black Duck is 'buudhanbang' in Wiradjuri and 'manparra' in Paakantyi; and so on, myriad times over.[12] Diversity of this kind runs counter to the nomenclatural standardisation that has long been a paramount goal of vernacular naming bodies. Choosing one Indigenous name from the plethora available for most species is a potentially fraught process: whichever name is chosen will privilege one Aboriginal language, leaving others out in the cold.

The Eungella Honeyeater is one of a very few species to which that dilemma probably did not apply. Being confined to the lands of a single language group, the Birri and Wiri peoples, in all likelihood it did not have a multiplicity of names in pre-colonial times. Unfortunately, however, the Birri and Wiri name for the Eungella Honeyeater has been lost, leaving a place name to carry the commemoration of the traditional owners of the lands on which the bird dwells. It's a more substantial Indigenous commemoration than most Australian bird names carry.

While the vernacular name, Eungella Honeyeater, has been stable since the bird was first described, over those four decades it's had three different scientific names. When Longmore and Boles named it *Meliphaga hindwoodi* in 1983, the genus to which they assigned it was under question. Richard Schodde had split *Meliphaga* into three genera, *Lichenostomus*, *Xanthotis*, and *Meliphaga*, in the mid-1970s. Had the Eungella Honeyeater then been known to science, Schodde would have placed it in *Lichenostomus* alongside its close relative, the Bridled Honeyeater. However, Schodde's generic split was still disputed among taxonomists in 1983, so Longmore and Boles took the conservative option and placed the new species in *Meliphaga*. By the 1990s, Schodde's three-way splitting of that genus had become generally accepted by taxonomists. So when Boles and fellow systematist Les Christidis published a new taxonomic list for Australia in 1994, the Eungella Honeyeater appeared as *Lichenostomus hindwoodi*.[13] But not for long.

In a major revision of the *Lichenostomus* honeyeaters in 2011, Árpád Nyári and Leo Joseph erected a new genus, *Bolemoreus*, into which the Eungella and Bridled Honeyeaters were bundled. The new genus name was a Latinised synthesis of the surnames of the two ornithologists, Boles and Longmore, who first described the Eungella Honeyeater.[14] That species, rechristened *Bolemoreus hindwoodi*, now bore both generic and specific names commemorating ornithologists. This is unusual. While it's not uncommon for a bird's specific or generic name to commemorate a person, it's rare for both names to do so. Among those rarities in Australia is the Gibberbird, *Ashbyia lovensis*, whose generic name honours the South Australian ornithologist Edwin Ashby, while its specific name memorialises the Reverend J.R.B. Love, a Presbyterian missionary who collected birds for Ashby. But *Bolemoreus hindwoodi* trumps that, commemorating not two but three people—and all ornithologists to boot! No other Australian bird's binomial is so bountifully commemorative.

The Eungella Honeyeater's rapid transition from *Meliphaga* to *Lichenostomus* to *Bolemoreus* suggests that we can expect the bird to be reclassified again soon. When this happens, Hindwood's memorialisation in the specific name will probably endure because, at least in the short term, the bird is more likely to be assigned to a different genus than for a reason to be found to rename the species. Still, you never know with these things, and avian taxonomy is undergoing one of its periodic upheavals, with a great deal of splitting—and a little lumping, too—under the impact of new techniques of genomic and DNA analysis. Scientific nomenclature has never been stable, but in the early 21st century it's in a phase of chronic instability.

Modern scientific nomenclature is not meant to be stable, but rather to reflect evolutionary history and taxonomic relationships. Systematists use the term 'phylogenetic tree' as a visual metaphor for those matters, but the metaphor is misleading because the science of phylogeny has none of the solidity of a tree. It's fluid and ever-changing, the various orders, families, genera, and species within Class Aves being continually reorganised, connected, disconnected, lumped, split, shuffled, and reshuffled. Because scientific understandings of avian evolutionary histories constantly change, there is no foreseeable end to scientific names changing, too. Indeed, recognition of the inherent instability of scientific names was among the factors that prompted ornithologists in the late 20th century to seek nomenclatural stability by standardising vernacular names.[15] It was a vain hope, perhaps, but in recent times vernaculars have been less volatile than scientific labels.

There's a long history of amateur birders scoffing at their professional counterparts' penchant for taxonomic revision and consequent name-changing. In a 1939 account of his birdwatching adventures, Charles Barrett concluded his description of a Metallic Starling colony on Hinchinbrook Island with a mock apology:

> *Calornis*, I wrote, forgetting that our shining starling should be *Aplonis*—*Aplonis metallica*. Systematists delight in name-changing; any excuse will do, though the Law of Priority is proclaimed from zoological house-tops. Well, it amuses them, I suppose. One of these experts, out in the bush with me, could not recognize birds whose old familiar names he had consigned to

the rubbish-bin for synonyms, giving new-old names in return.
To him, its scientific title was more than the bird.[16]

Barrett wrote this at a time when Gregory Mathews was still active and the chaos into which he had plunged Australian avian taxonomy still rankled with many birders. Decades later, long after the dust had settled on Mathews' mayhem, many birders were still bemused by the turmoils of taxonomy. As Graham Pizzey wrote in the first edition of his field guide, 'The classification of birds (avian taxonomy) is in some ways like the peace of God—it passeth all understanding.'[17]

Yet while birdwatchers might scoff at what they regard as taxonomists' follies, they still defer to the taxonomists' decrees. Pizzey arranged the birds in his guide in the scientifically accepted taxonomic order of the day, and didn't dare deviate from each species' recognised scientific name, although he happily refused to follow the RAOU's recommendations on vernaculars. Most recreational birders in recent times haven't bothered to learn the scientific names of most species, although the further back in time one goes, the less valid that generalisation becomes. With or without their Latinised labels, however, the categories erected by systematists are honoured by birdwatchers. More than honoured, they're revered.

The category of species carries a special cachet. It's the foundational unit of birdwatching. Differentiating species is the principal activity in birdwatching excursions; seeing new species is a driving ambition; adding a tick to a species list is an occasion for rejoicing. Sometimes, birders may turn to identifying subspecies or listing genera. But species are at the heart of the pastime. And whatever misgivings recreational birders might have about the instabilities of avian classification, they still respect the foundational status of species as a category, and abide by taxonomists' verdicts on where the dividing lines between species should be drawn. The fact that the taxonomists themselves may disagree on the latter issue allows a level of leeway, but the point remains that recreational birders draw their crucial criteria from an external source: science, the most authoritative arbiter of truth in the modern world. That external arbiter adds to the pastime's strength.

Birders have always deferred to taxonomists. In his pioneering field guide of 1911, John Leach proclaimed his adherence to the classification system of the distinguished British ornithologist Richard Bowdler Sharpe. At one point, he expressed faint reservations about some of Sharpe's determinations, but added, 'However, Sharpe's classification represents the latest thoughts of scientists on this difficult matter, so it must be adopted here.'[18] Later field guide authors, such as Pizzey, followed Leach's example. Scientific taxonomy had to be adhered to, lest birding collapse into the chaos lamented in the Bible when 'there was no king in Israel [and] every man did that which was right in his own eyes'. No modern, mass-participation recreation can abide such licence.

Even the most fanatical twitchers defer to science on what they can tick. They may grumble when taxonomists erase a treasured tick by lumping two hitherto separate species into one. They may rejoice when scientists, by splitting one species into two, allow them to add another tick to their lists without exerting the least effort. In twitching parlance, there's even a term for this phenomenon: 'armchair tick'. But no twitcher would dare conjure an armchair tick by splitting a species themselves. That, by unspoken consensus, is the exclusive preserve of the scientific taxonomist. The most choice that a twitcher—or a recreational birdwatcher at any level of enthusiasm—can exercise is between the two or three scientific taxonomies on offer at any one time, and they typically differ in only minor respects. By its deferral to scientific order and adherence to consensual rules, birdwatching shows its origins as a pastime born in modernity.

No one got armchair ticks out of the reclassifications of the Eungella Honeyeater, because they involved reassignments to new genera rather than the creation of new species. However, J.S. Robertson, F.M. Hamilton, and Alec Chisholm, who thought that what they saw in 1959 and 1964 were Bridled Honeyeaters, could have been awarded armchair ticks in 1983 (when the Eungella Honeyeater was split from the foregoing), had they been still alive and interested in the sport of competitive ticking. They weren't, on either count.

CHAPTER SEVENTEEN

Common Myna: Foreigners

'CANE TOADS WITH WINGS' headlined a 1994 *Wingspan* article on the Common Myna by RAOU conservation officer Hugo Phillipps. Traducing the species as the 'scavengers of suburbia, the bullies of the boulevards, the masked bandits of the metropolis', he warned that, 'Only when we are all aware of the menace behind the mask will we be able to control the Common Myna.' By 'control', he meant eradicate wherever possible.[1] A few years later in the same magazine, CSIRO ecologist Dean Graetz introduced himself as a man who 'enjoys the presence of Australian birds just for pleasure', immediately adding that he was 'interested in devising an effective and relentless strategy for killing Indian (Common) Mynas'.[2]

By the 1990s, animosity toward mynas—including the explicit expression of a craving to kill them—was commonplace among birders and biologists. But it did not go uncontested. The 'Cane Toads with Wings' article prompted a response from E.L. Jones, who deplored 'Mr Hugo Phillipps' plea for ethnic cleansing with respect to Mynas'. This was the time of the Yugoslav Wars, and 'ethnic cleansing' had recently entered the lexicon as the term for an exceptionally vicious kind of atrocity driven by an excess of nationalist fervour. Persisting with that link, Jones argued that, 'It would be unfortunate if public funds and zoologically trained people were to try to exterminate Mynas for what are political rather than incontestable aesthetic or economic reasons, let alone scientific ones.'[3] The next issue of *Wingspan* carried another response by ornithologist Heather Gibbs. 'I used to despise Mynas', she admitted, 'but now believe scientific research is crucial to their cost-effective and ethical management and would like to encourage people to reassess their attitude towards Mynas.' 'Adequate knowledge [of myna behaviour and ecology] does not yet exist', she pointed out, with an implication that antagonism to the species was driven, at least partly, by sheer prejudice.[4]

Australian birders' attitudes to introduced species have varied, but there's been a discernible increase of antipathy since the turn of the 20th century, with hostility escalating steeply in the last 50 years or so. Growing hostility was doubtless connected with a growing ecological consciousness and awareness of the environmental damage caused by exotics. But more was involved. The heart-felt hatred of introduced species, voiced by many birders, draws on something deeper than mere awareness of ecological interrelationships. It draws on a conviction that these birds—mynas and sparrows and starlings, and so forth—have no legitimate place here. They don't belong. Intensifying antagonism toward introduced species correlated with a growing devotion to nativeness as a precondition of biological worth. As several historians and anthropologists have pointed out, recent environmentalist advocacy often conflates the concepts of nature, native, and nation, a conflation that is particularly evident in projects to restore the national territory (or selected portions of it) to a supposed state of nature by cleansing it of all non-native (aka 'alien' or 'invasive') species.[5]

That ambition began to be pursued in Australia in the last quarter of the 20th century, a time when animosity toward introduced birds intensified. Heralding the upsurge of enmity, Ken Simpson penned an article in January 1975 on 'the greatest threats to birdlife', in which, referring to introduced species, he told readers to 'steel your hearts ... against the invasion of an army of feral and foreign nationals every bit as lethal as an armed military invasion of Australia'. Lumping together 'starlings, Indian turtle doves, Indian mynahs, Sparrows, Blackbirds, Spice finches, Bulbuls, Skylarks, Rock pigeons, Mallard' with 'cane toads, cats and dogs, rats and mice, rabbits, foxes, ferrets', Simpson urged people 'to get out there and eradicate them with every weapon at hand'. In an inept extension of the military metaphor, he declared that, 'We should not sit by and watch a series of floral and faunal Dunkirks.' More straightforwardly, he denounced all introduced species: 'We don't want them; they don't belong here.'[6]

Perhaps there was an element of theatricality in Simpson's declamations. Yet it's noteworthy that while his article covered the gamut of threats to birdlife—including pollution, habitat loss, and 'eco-system modification'—it was only on introduced species that he resorted to military metaphors and emotive exhortations. On other threats, his exposition was more restrained. The other threats were amorphous, insidious, or intangible, whereas introduced species were real, living, breathing creatures. Living creatures can arouse feelings of love and amity, but equally they can incite loathing and enmity. Birders' emotional responsiveness to birds cuts both ways.

Not everyone was so dramatic in their denunciations, but there was a change of tone around the time Simpson wrote. Earlier birders had expressed misgivings about introduced species, but they had not vented such vehement and violent hostility as Simpson and many of his successors did. In earlier times, dislike tended to be counterbalanced with some measure of affection; disapproval with acceptance, albeit often grudging. Moreover, in the early 20th century there was a residual respect for birds that, only a short time earlier, had deliberately been brought to these shores.

Most avian introductions to Australia were made in the second half of the 19th century, with 'usefulness' usually a criterion for a species' selection. It was partly for their usefulness as insect-eaters that Common Mynas were introduced to Melbourne in the 1860s; and their pest-control function was an even stronger motive behind their release in the north Queensland cane-fields in the 1880s. Yet no species was introduced solely for utilitarian reasons. Aesthetic and sentimental considerations were always in play. That was the case for the Common Myna, and its usefulness was soon downgraded.

Robert Hall had an entry for the Indian Myna in his 1907 book, *The Useful Birds of Southern Australia*, but it made no mention of the species' usefulness. Instead, it applauded the myna on aesthetic grounds, as a bird that 'makes cheerful the environment of the cities' and 'adds considerably to the joy of living'.[7] Walter Froggatt's *Some Useful Australian Birds* of 1921 had no entry for the myna, its only mention being a brief one under Noisy Miner, asserting the need to distinguish the latter 'from the very different Indian Minah, which is somewhat of a house pest in some of our towns'. Froggatt noted that introduced birds in general caused 'serious problems', and indicated that the role of pest-controller was best left to natives.[8] Lucas and Le Souëf in 1911 took a similar line. Acknowledging that some introduced species ate insect pests and in other ways helped the farmer, they added, 'We have large numbers of indigenous birds which perform the same good offices', so the wisest course of action was to protect and preserve the native species.[9]

Early-20th-century birders asserted the usefulness of native birds far more frequently and fulsomely than they did for introduced species. Insofar as the usefulness of introduced birds was mentioned, it was usually as a brief aside; often adding a note that native species were equally, if not more, useful; and often, too, with an acknowledgement that some introductions had transmuted from pest-controller to pest. Already, there was concern about exotic birds' impact on natives.[10] However, the term 'pest' was usually reserved for those that adversely affected human economic enterprises or inconvenienced people in other ways—for example, by spreading lice. House Sparrows and Common Starlings were among those most vehemently vilified as pests, whereas Eurasian Skylarks and Song Thrushes were seldom so labelled since their numbers were few and their economic demerits negligible. Besides, they sang gloriously. Common Mynas seem usually to have been positioned somewhere between those two extremes: a bit of a pest sometimes, but with pep and personality. In any event, it was anthropocentric considerations that earned the 'pest' tag.

In a path-breaking paper on 'the menace of acclimatization', presented to the 1936 RAOU congress, Dom Serventy critiqued such anthropocentric assumptions. We were mistaken, he argued, in assessing certain introductions as 'neutral or harmless' merely because they did not directly damage human interests. That such assessments were made by people 'who profess regard for the native species'—a not-very-veiled reference to his RAOU audience—was particularly 'disquieting'. What must be understood, Serventy insisted, was 'that *any* successful introduction, even if directly innocuous from the human standpoint, must, by very reason of the fact that it has obtained a foothold, disturb the balance which had existed, and therefore have repercussions which will detrimentally affect the existing fauna'.[11] This was not Australia's first airing of ecological arguments against introductions, but Serventy put the ecological case with exceptional eloquence and cogency.

Serventy's 1936 RAOU presentation was prompted by gun-clubs lobbying for the release of foreign game birds. Six years later, more ominous circumstances provoked him into again warning against the dangers of avian aliens. It was early 1942: fears of Japanese invasion were at their peak, and Australians were anxiously discussing how best to evacuate parts of the country in the event of enemy incursions. A veterinarian, Geof Fethers, recommended euthanising pets that could not readily be turned loose, but setting cage birds free. Serventy strongly advised otherwise. The chances of cage birds surviving in the wild were slender, he admitted, but the risks were far too great. If released birds did manage to survive, it could be only 'at the expense of some native bird', so cage birds, too, if they could not be taken to a place of safety, 'should be destroyed, and not freed'. This might sound 'drastic', he admitted, but for the sake of our native avifauna it was necessary.[12]

Birders listened to Dom Serventy. He was arguably Australia's most highly regarded ornithologist in the middle decades of the 20th century, with a knack for communicating across the amateur–professional divide. Delegates at the 1936 congress concurred with his arguments, and passed a motion condemning floral and faunal introductions.[13] But an ecological outlook was only beginning to germinate, and for decades afterwards birders happily voiced their affection for introduced species—or, at least, some such species, Common Myna among them—with barely a nod toward the ecological implications of their presence. Besides, Serventy's arguments were against future introductions. On what could or should be done with those already here, he was silent.

In the mid-20th century, there was nothing unusual about eminent birders extolling exotics. Charles Bryant, editor of the *Emu*, published a paean to the Song Thrush in 1944. It was primarily an aesthetic appreciation of 'this avian musician of the suburbs', although he also complimented the bird for helping control garden snails. Bryant gave no hint that the thrush was out of place in Melbourne, writing instead as if it rightfully belonged there.[14] A decade later, broadcaster and university-trained biologist Crosbie Morrison publicised his affection for the Common Myna, a bird 'with a rather attractive swagger'. The myna, he urged, should be welcomed in the same way as Australia was then welcoming its increasing intake of human migrants, for 'he is a New Australian worth assimilating'.[15] Bryant and Morrison were well aware of ecological arguments of the kind advanced by Serventy. On other occasions, they advanced similar arguments themselves. But ecology had not yet become ideology, and birders then had few inhibitions about publicising their emotional appreciations of birds.

For emotionally appreciating birds, no one outdid Alec Chisholm. Accordingly, his attitudes toward introduced birds were ambivalent, combining fondness for them with an overt preference for natives. The Common Myna merited 'esteem for its heartiness and self-confidence', he wrote in 1964, but added that as a 'personal preference' he 'would readily exchange a whole batch of mynas, or of any other feathered imports, for a single pair of two small natives', the Scarlet Honeyeater and Jacky Winter.[16] In an earlier item, he labelled the predominance of introduced species' in Melbourne's avifauna a 'misfortune', but continued:

> Nevertheless ... I doubt if I would be willing to lose certain introduced birds from Victoria's capital city—the Myna with his important strutting about roadways and his animated chatter, chuckle and song; the Blackbird with his Celtic melody; and, above all, the Thrush with his quiet ways and charming song. These three birds, together with the little native Chickowee, remain for me part of the Spirit of Melbourne.

Despite their foreign origins, Chisholm judged the myna, Blackbird, and Song Thrush to belong in Melbourne's cityscape as fittingly as the Mangrove Gerygone did in Brisbane's and the Jacky Winter in Sydney's.[17] He had a special fondness for the Common Myna, writing in 1971 that this bird 'is now part of the Australian landscape. And it is not, I suggest, an objectionable feature'.[18]

Chisholm's bird preferences were informed more by aesthetics than ecology. They were infused, too, with the nature-based nationalism he tried to foster in his fellow Australians. His love of birds could transcend the native–alien divide, but the bridging was partial, in both senses of the word. Believing that 'imported birds rob a landscape of its national character', he was perturbed by places in which exotics outnumbered natives and in which they had established a strong presence outside urban settings. Yet he accepted—even celebrated—the presence of introduced species in limited numbers and circumscribed spaces.[19] Similar attitudes were common among mid-century birders.

From the 1970s, attitudes stiffened into stronger and more sweeping animosity, even violent enmity of the kind voiced by Ken Simpson. This may have correlated with introduced species expanding in numbers and range around this time. The Common Myna did so. But at least one exotic, the House Sparrow, declined sharply in the late 20th century without any diminution of birders' detestation of them. Hardening attitudes were products of human culture rather than avian numbers.

A hardening of attitude toward introduced species was widespread, but not uniform. Some birders—perhaps many—retained touches of ambivalence and affection. In 1977, Ted Schurmann wrote of introduced birds in a generally though gently negative tone, leavened with praise for some such as Goldfinches, Skylarks, and Song Thrushes for their attractiveness and pleasant songs.[20] Peter Trusler,

Tess Kloot, and Ellen McCulloch in their *Birds of Australian Gardens* (1980) acknowledged that introduced species could be 'nuisances' and compete with native birds, but conveyed an overall attitude of acceptance. Although acceptance could sometimes be grudging, they nonetheless delighted in the melody of the Song Thrush, the 'bright colours, tinkling calls and quick, light movements' of the Goldfinch, and the 'larrikin' behaviour of the street-smart Common Myna (which Trusler delightfully depicted peering at a crumpled Cheezels packet in a gutter).[21] In his 1985 book *Garden Birds*, ornithologist Clifford Frith discussed introduced species in a usually neutral, sometimes appreciative, tone. 'It is true', he noted, that the Common Myna 'occupies nesting cavities in buildings and trees that native birds might otherwise utilize. It is equally true, however, that this is a handsome and enjoyable bird to look at.'[22]

In the 1980s, Rosemary Balmford went further than most on behalf of introduced species. She urged Australians to appreciate and admire them, although the tone of her pleading shows she knew she was swimming against the tide. In a 1988 issue of the *Bird Observer* she expressed a hope that Australians were 'sufficiently mature, and sure of ourselves in our own country, to accept, and to enjoy, reality':

> Because, after all, if you enjoy birds, if you love birds, if you find birds beautiful or interesting, if birds add a dimension to your life, if you find yourself travelling in order to see birds—what difference does it make to you where a bird's ancestors originated? ...
>
> We brought them to this country. Like ourselves, they are here to stay. And no one who has heard the bird that "singing still dost soar and soaring every [sic] singest" could regret the introduction to Australia of the Skylark.[23]

Quotations from Shelley fell flat for many readers. Repudiating Balmford's pleas, Glen Johnson told readers of the *Bird Observer* that all introduced species were 'vermin' and 'should be treated accordingly'.[24]

Vilification as vermin triumphed over panegyrics in verse. As the turn of the 21st century approached, introduced species were increasingly demonised, albeit with ecological sanction. The Common Myna, once imbued with cheeky charm, became the most despised bird in Australia. A contributor to the *Canberra Times* in 1990 described it as 'a kind of feathered cockroach'.[25] In the same city in 2006, an Indian Myna Action Group was founded, with the sole objective of killing the birds. Similar bodies have since been established in numerous other towns and cities.[26] Facilitating their actions, Common Mynas, unlike native species, are not protected in any Australian state or territory. Killing must be done 'humanely', but that's an inevitable proviso in a modern Western nation. The facts that killing is an open option for alien avifauna, but not for natives, and that the killing is promoted as an act of environmental restoration, carries more than a whisper of nativist enthusiasm.

Canberra birder Noel Luff concluded a 2016 article on attitudes toward the species by asserting that, 'There is now almost universal agreement that the Common Myna is indeed a pest in Australia. I doubt if one could find anyone who would be prepared to put in a kind word for it.'[27] The first sentence is undoubtedly correct, but the second does not follow from it. Numerous birds and animals are recognised as ecological pests—cats and dogs, Noisy and Bell Miners, to name just a few—but that does not mean they are regarded as nothing but pests, and that no one has anything positive to say about them. Our human responses to other life forms can be more accommodating than that—they should be, since as ecological pests we humans far outdo all others. As a bird lover as well as a birdwatcher, I'm happy to put in a kind word for the Common Myna, even while I recognise that it's a pest. I suspect that's what the old-school birders were getting at.

CHAPTER EIGHTEEN

Rainbow Lorikeet: Feeding

IN A 1910 *EMU* article, Mackay RAOU member Edward Cornwall delightedly described a neighbour, Mrs Alex Innes, 'engaged in her daily pleasurable task of feeding a number of Blue-bellied Lorikeets' (as Rainbow Lorikeets were then called). 'When the lady calls', he enthused, 'they come by the dozen for the food she offers them, and settle all over her shoulders, head, hands, &c, and on the table, which is placed on the verandah for their special benefit.' An accompanying photograph illustrated the joyful mayhem. Cornwall delighted not only in the colourful spectacle, but also in the fact that the lorikeets that Mrs Innes loved had 'learned to trust her implicitly'.[1] As birders of his generation knew, building trusting and affectionate relationships with birds was the point of feeding them.

Later generations of birders knew that, too. The April 1947 issue of *Wild Life* magazine carried a charming photograph of a little girl, one arm outstretched to four Apostlebirds feeding from her hand. It's captioned 'Friendship'.[2] An article in an earlier issue of the same magazine elaborated on the point at some length. Frank Robinson's

Figure 19. This full-page spread in the April 1947 issue of *Wild Life* was simply captioned 'Friendship'.

'In Favour of a Bird Table' began by urging 'friendship with birds', and concluded by observing that bird-feeding is 'something of a social exercise' linking human and avian participants. In between, Robinson explained that, when 'feeding birds ... I feel like an honoured guest, privileged above my fellows because the bird will come to my table and eat in my presence without fear. That, then, to me is the virtue of a bird table. Its effect is not to train the birds, but to train human beings'.[3]

Three decades later, bird-feeding and the friendships it fostered still got the tick of approval from birdwatchers and ornithologists. In his 1977 primer on *Bird-watching in Australia*, Ted Schurmann advised that, 'If you want to be on really close terms with your birds, try feeding them.' Around the same time, the Bird Observers' Club promotions officer, Ellen McCulloch, stated that 'feeding birds can be just about the most powerful public relations exercise you can make', since it encouraged people to connect emotionally with the birds around them. Both Schurmann and McCulloch supplemented their advocacy of feeding with recipes for bird puddings and other avian foods.

They also sounded a note of caution. While he advocated feeding, Schurmann warned of 'the danger of over-feeding and unnecessarily prolonged feeding'. McCulloch, too, cautioned against allowing birds to become 'too dependent' on handouts, and acknowledged that 'some people feel concern at this practice' of bird feeding. She did not share their concern, provided that feeding was done appropriately and hygienically, but her words show that murmurings of disquiet were being voiced in the late 1970s.[4]

Over subsequent decades, those murmurings became louder and louder, eventually escalating into howls of condemnation. In a typical recent denunciation, a Sydney Wildlife Rescue website asserts that it is 'cruel to feed our native birds'.[5] The New South Wales government even hosts a website explaining 'How can I stop my neighbour feeding the birds?' It claims that 'most people feed birds out of a misguided desire to care for them', but its real consequences are 'health problems or even death for the birds', following with helpful hints on how to persuade your neighbours to desist from the dreadful practice.[6]

Condemnation now dominates official discourse on bird-feeding in Australia. It's a scenario unique to this country, as ecologist Darryl Jones has pointed out in several recent books. In Europe and North America, by contrast, bird-feeding carries the blessing of bird lovers who, far from trying to stamp out the practice, do all they can to promote it. Jones is at the forefront of a push to change the dominant narrative on bird-feeding in Australia to align with that overseas.[7] To judge from the publicity he and other proponents of bird-feeding have been given, that message finds an appreciative audience in Australia.

While Jones clearly explains that the dominant stance against bird-feeding in Australia contrasts with the mainstream view overseas, he misses the important point that Australia's anti-feeding stance only recently gained ascendancy. Misleadingly, he states that:

> Until very recently there was an incontestably obvious presumption that people don't feed birds in this country. Among ecologists and conservationists, birdwatchers and environmentalists, it was practically an article of faith, an accepted dictum: we all know we shouldn't and therefore we don't.[8]

In fact, until recently the predominant presumption was that feeding birds is self-evidently good. Throughout most of the 20th century, Australian birders enthusiastically promoted it.

Probably no promoter was more enthusiastic than the Gould League of Bird Lovers, which from its inception in 1909 encouraged bird-feeding as part of its efforts to foster a conservationist consciousness in children. 'A bird feeding table is an essential part of every bird lover's garden', Gould League publications proclaimed.[9] More than merely advocating feeding, league publications gave recipes for bird puddings and other avian eatables. For sugar syrups, they gave not only recipes, but also instructions on how to construct devices from which birds could slurp up the sweet liquid. Alternatively, taking less trouble, 'Scraps of food may be placed on a feeding-table for the birds—sugar and honey for the honey-eaters, fragments of cheese and meat for the insect eaters.'[10] League publicity persistently flaunted photographs of smiling children feeding birds, sometimes from their own hands, sometimes from the bird tables it encouraged all schools to erect.

Writing in 1927, Melbourne RAOU member J.D. Jennings lauded Bird Days for having 'created a "bird-conscience" in our boys'. In his own boyhood, he recalled, a boy won esteem from his peers according to how many birds he killed with his shanghai, but:

> Now, what a contrast we have! He is accounted great who can invite his mates to his house and, at his magic call, bring wild birds to partake of his hospitality—to perch upon his arm and eat from his palm—to hold a worm, the while a yellow robin breaks dainty pieces from it.

Feeding birds was a wonderful way to earn their trust, Jennings enthused, and the human-avian bonds thus created could only benefit our birdlife.[11]

Feeding birds wasn't just for the kids. Adults participated, too, usually for the sheer joy of interacting with wild birds, but occasionally with ulterior motives as well. John Rowland Skemp's 'My friend Midgie' exemplifies one such mix of motives. Skemp was a Tasmanian naturalist who wrote amiable tales about the birds he befriended on his farm near Launceston around the middle of the 20th century. In 'My Friend Midgie' he told 'the true story of a little Blue Wren' whose trust he won, first with cheese crumbs scattered on the windowsill, then with termites that Midgie quickly plucked up the courage to eat out of Skemp's hand. There's 'a pleasant little thrill', he confided, in 'having these tiny wild things eating trustfully from one's hand.' Throughout the story, Skemp used food, first to make connection with wild birds, and thereafter to consolidate the friendships. He also used the food-forged friendships to facilitate close observation of birds' behaviour and appearance.[12]

The original version of 'My Friend Midgie' was published in a 1949 issue of *Wild Life*. A later version, in a collection of Skemp's writings published after his death, added an ornithological postscript. It was the last thing he ever wrote, scrawled on a scrap of paper while he lay dying in the Launceston General Hospital in March 1966. In it, Skemp set out his observatohns on annual plumage changes in male Superb Fairy-wrens, giving meticulous detail on when they changed from eclipse into breeding plumage, and even on the deposition of pigmentation on the barbules of the feathers. It was the kind of detail that ornithologists usually derived from dead specimens, but Skemp made his observations on living birds that had perched trustingly on his fingers to feed.[13]

Feeding was also used to facilitate bird photography, especially in the days when cameras were cumbersome contraptions. In their pioneering 1920 book on bird photography, for example, Littlejohns and Lawrence explained how they lured kookaburras to the camera by nailing chunks of steak to a stump. This, they admitted, was 'the nearest approach to picturing a tame bird of which we have been guilty', for the kookaburras were accustomed to being fed by local residents.[14] A 1948 *Wild Life* article by Lucy Lee recounted the pleasure she took in feeding, both 'as a gesture of friendship between us and the birds' and as a handy way of getting them to pose for a photograph. Crosbie Morrison added an editor's postscript explaining that feeding was 'one of the most productive ruses in bird photography' and the only reliable alternative to finding them at nest.[15] Similarly, in his 1955 *Handbook of Elementary Bird Study*, Patrick Bourke indicated that while the nest was the usual place to photograph birds, food trays could serve as ready-made 'studios'.[16]

In her 1980 book, *Learning About Australian Birds*, Rosemary Balmford recommended 'baiting with appropriate food' as a great way to get bird photographs. Like Littlejohns and Lawrence 60 years earlier, she suggested nailing a lump of meat to a stump as an excellent ruse to secure photographs of kookaburras. 'When you have got your picture, give the bird the meat', she added. The revised 1990 edition of Balmford's book gave the same advice.[17]

Most of the time, however, birders fed birds not for any ulterior motive, but for the sheer joy of interacting with wild creatures. And most mentions of feeding in the birding literature, at least until the 1990s, carry no hint of a need to justify the practice. Many were simply observations made in passing, as if feeding was such an unremarkable activity as barely to warrant comment. In her charming 1950s story about the 'Birds of my garden', for example, Ada Fletcher mentioned feeding merely as part of her pleasant daily routines of bird appreciation.[18] In the *Bird Observer* at this time and for decades afterward, numerous articles made fleeting mentions of bird-feeding as if it was so common as to deserve no special notice.

There were, however, enthusiasts who promoted feeding with zeal. Barbara Salter was one. In innumerable magazine articles and in her 1969 book, *Australian Native Gardens and Birds*, she not only urged people to feed birds, but also gave recipes for bird puddings, porridges, syrups, seed cakes, and other avian delectables.[19] Sometimes, her recipes caused dissension. At the end of 1969, for example, Salter disputed with fellow bird-feeder Florence Vasey over whether artificial nectar for honeyeaters should be based on white sugar or a brown sugar-and-honey mix.[20] At that time, there were far more disputes over what to feed the birds than whether to feed them.

Before the 1970s, outright opposition to bird feeding was very seldom voiced in Australia. Even in that decade, there was little opposition although cautions were sometimes sounded, as if the self-evident rightness of the practice was beginning to be questioned. Into the 1980s, the questioning intensified.

In 1985, Ellen McCulloch noted that opinions on feeding 'range from that of the purists who are dead set against any provision of extra food anywhere for any birds whatever the circumstances, to those who festoon their gardens with permanent food fore and aft, and have to get in a bird-sitter before they go on holidays'. Evidently, disagreement had developed, but the 'purists' were not yet ascendant. McCulloch herself acknowledged that she fed birds in her garden; she encouraged others to do so, and illustrated her article with a photograph of two honeyeaters at a feeder of the kind sold at the BOC shop.[21]

Five years earlier, McCulloch had collaborated with Tess Kloot and artist Peter Trusler to produce *Birds of Australian Gardens*. Feeding birds will add 'interest and pleasure to your life', they assured readers, adding helpful advice on what to feed, how to make bird puddings and artificial nectars, and how best to get these foods to the birds. Aware that others held misgivings about feeding, they justified the practice as a means 'to redress the balance of nature' in urban areas, at the same time warning against allowing birds to become 'too dependent on an artificial food supply'.[22] Graham Pizzey also warned against fostering dependency in his 1988 book, *A Garden of Birds*. But he made no excuses for his advocacy of feeding, giving recipes for sugar syrup and bird pudding, plus advice on how to dispense these foods, without a hint that such recommendations might provoke controversy.[23]

In the 1990s, opposition to feeding stiffened. Even so, opposition was usually less than total. In their 1998 birdwatching manual, Ken Simpson and Zoë Wilson noted that there was 'some debate over the matter of feeding birds in the garden. Some people see no harm in it, others regard it as an immoral act'. Simpson and Wilson did not go to the latter extreme, although they did express misgivings about feeding, and clearly preferred the supposedly more 'natural' option of planting bird-attracting trees and shrubs. Accepting that some level of feeding would occur anyway, they counselled 'moderation', and warned against birds becoming 'dependent' on human-provided food. In one of their more morally charged sentences, they declared that, 'We're talking about the pleasures of watching wild birds, not encouraging overweight suburban freeloaders.' Still, they allowed a space for feeding, albeit hedged around with caveats and provisos. They even included a photograph of a Crimson Rosella eating sunflower seeds from a human hand while the other hand held a field guide open at the plate on parrots. It was captioned 'The ultimate bird identification experience'.[24]

It is impossible to pin-point a time when the anti-feeding lobby won ascendancy in Australia. For one thing, it happened differently for different groups, with some interested parties such as environmental bureaucrats and ecologists taking an anti-feeding stance sooner and stronger than others. Perhaps that's why ecologist Darryl Jones—who admits he was 'in the "Never feed!" camp' before he investigated the issue[25]—exaggerates the ubiquity of the 'never feed' message. He's right to point out the strength and salience of that message in recent times, but its ascendancy was never total. Among the groups with a special interest in the matter, birdwatchers maintained a measure of ambivalence.

For example, a *Wingspan* article ushering in the new millennium listed '50 things you can do to help birds, birding and Birds Australia'. Number seven advised that the 'best source of food for native birds are Australian plants, but if you want to feed birds in your garden do so on an ad hoc rather than regular basis so that they don't become too dependent'.[26] Eight years later in the same magazine, Michelle Plant published an article on 'Good practice when feeding wild birds', advising on both the risks associated with the practice and how to overcome them.[27] Darryl Jones himself had an item in the same issue, rehearsing the arguments he later elaborated in *The Birds at My Table*.[28] When David Andrew published *Pocket Garden Birdwatch* in 2015, he threw caution to the winds and advocated feeding unreservedly. Feeders 'are a must for any bird garden', he enthused, recommending black sunflower seeds as 'an excellent year-round food', and rice, whether brown or white, as 'a good choice for birds'.[29]

That said, there's no doubt that feeding became increasingly controversial and opposition to it increasingly prevalent among birders from the late 20th century onward. Why that was so, uniquely in Australia, is a puzzle I've not yet unravelled. Perhaps it owes something to the growth of a purist conception of nature associated with the wilderness movement that burgeoned in Australia from the 1970s onward. Its devotees decoupled humanity from nature, celebrating as 'wilderness' those patches of the planet that supposedly had evaded the withering touch of humankind. Anecdotally, this correlates with views I've heard from opponents of feeding, who suggest that it somehow violates the naturalness of birds, reducing them to the playthings of people. However, the wilderness movement was not unique to Australia. It was Western-world-wide, arguably reaching its apogee in the US. But the rest of the world did not experience a backlash against bird-feeding, and Americans kept happily feeding chickadees and nuthatches while Australians soured against doing the same for honeyeaters and lorikeets.

Yet even in Australia, the anti-feeding lobby had limited success where it mattered most: in backyards. Early in his research, Jones found that feeding birds remained popular among the wider populace, despite the injunctions against it from above. Amazed by this, he investigated why so many people persisted in feeding birds, uncovering a multitude of motives behind its popularity. Prominent among them was a desire to connect with nature.[30]

As in birdwatching, communing with nature isn't the only impetus behind feeding, but it's a powerful one, and always has been. It was evident in Mrs Innes feeding the Rainbow Lorikeets in her Mackay garden in 1910, and in the Gould League's many celebrations of children reaching out to birds with food in their hands and love in their hearts. It was evident in the 'pleasant little thrill' that John Rowland Skemp felt when Midgie fed from his fingers, and in Barbara Salter's delight in preparing puddings for her backyard birds. It's still evident in innumerable instances of birders and non-birders alike delighting in bonding with birds by sharing food with them.

In their 2016 book, *Visions of Wildness*, Peter Slater, Raoul Slater, and Sally Elmer extoll the wonders of the wild—or 'wildness' as they call it, as distinct from 'wilderness'. Wildness is the world of everyday nature, and, as Raoul tells us, 'wildness comes to those who are open to its embraces'. The book's final chapter, 'Wildness in urbia' encourages that embrace in urban settings, including by feeding birds. 'Purists decry the feeding of wild birds', the authors acknowledge, 'but more benign observers see it as maintaining a link with nature.' Those words are illustrated with photos of Rainbow Lorikeets 'filling up on handouts at a back door', colourfully attesting to the authors' point that embracing the wild is accessible to us all.[31] Feeding birds is one way of experiencing that embrace. It might help nourish the birds, but it's more to nourish our souls.

CHAPTER NINETEEN

Golden-shouldered Parrot: Saving

UNLIKE ITS COUSIN, THE Paradise Parrot, the Golden-shouldered Parrot has side-stepped extinction. But it's been a close-run thing.

The two species are very similar in appearance—so similar that the first European person to see a Golden-shouldered Parrot thought it was a Paradise Parrot. John Gilbert was the observer, and he knew the latter species well, having discovered it on the Darling Downs in 1844. A year later, as a member of Ludwig Leichhardt's expedition to Port Essington, he saw flocks of parrots on the Mitchell River in southern Cape York Peninsula that he identified as the same species. Gilbert was an expert field ornithologist, and he probably made this mistaken identification because he didn't do what then had to be done to guarantee accuracy: he didn't shoot the bird, but merely looked at it, probably from a considerable distance.

Golden-shouldered Parrots resemble Paradise Parrots not only in appearance, but in habits and habitat requirements as well. Both are birds of open, grassy woodlands, where they feed on a seasonally varying array of seeds. Both build their nests by tunnelling into termite mounds, excavating a chamber in which they lay five or six white eggs. Like its golden-shouldered relative, the Paradise Parrot probably relied on the excreta-eating larvae of a moth to maintain nest hygiene. Both birds are ecologically highly specialised.

The first specimens of Golden-shouldered Parrots were taken in 1855 by John Elsey on Clarina Creek, east of the present-day town of Normanton. There are no Golden-shouldered Parrots there today, because the species is one of the many that has suffered from European colonisation, contracting in both numbers and range. Their decline began when European pastoralists brought their cattle and the environmental-management strategies of open-range grazing onto the parrot's lands. That process began in the 1870s, and the subsequent history of the Golden-shouldered Parrot eerily echoes that of its close congener, the Paradise Parrot—with one big difference.

The Golden-shouldered Parrot's saving grace was that Australia's northern lands were not as comprehensively colonised as those in the south.[1] On Cape York Peninsula, the European population remained small and scattered; the pastoral industry spread slowly and was always economically precarious; intensive agriculture was never established; European-induced environmental impacts, although considerable, were of a different order of magnitude from those down south; and the Indigenous peoples remained both numerous and, to a large extent, on or near country. These would be crucial factors behind the survival of the Golden-shouldered Parrot. Similar circumstances allowed the related Hooded Parrot of the Northern Territory to continue to thrive, while the Paradise Parrot of inland south-east Queensland fell fitfully to extinction.

According to the ornithologists who have studied the species most closely, Gabriel Crowley and Stephen Garnett, Golden-shouldered Parrots were abundant across western and central Cape

York Peninsula until the early 20th century. Collector William McLennan and ornithologist Donald Thomson, working separately in the Coen district, found the parrots to be reasonably numerous there in the 1920s. By the 1950s and 1960s, their decline was well underway, although the parrot population was still adequate to sustain an illegal industry of trapping for the aviary trade. This deepened the parrot's plight. By the 1970s, its range was only about half what it had been 100 years earlier, and since then has shrunk to two small patches: one centred on the Morehead River, and the other on the Staaten River.[2] It was a slower diminution than the Paradise Parrot's, but still a steady slide downhill.

When Crowley and Garnett began their research in 1992, the Golden-shouldered Parrot was 'one of Australia's most threatened birds' with a wild population estimated at fewer than 1,000. The precise nature of the threats was unknown, and information on the parrot's habitat requirements, movements, and breeding biology was sketchy. With funding from the Queensland Department of Environment and Heritage and the World Wide Fund for Nature, Crowley and Garnett set out to change that, ascertaining why the species had declined and how to arrest—and hopefully reverse—the decline. Basing themselves at Sue and Tom Shephard's Artemis Station near Musgrave, they managed to find 'more of the birds than we expected', and soon began unravelling some of the intricacies of the parrot's ecology. But the area they had to cover was vast, and as just two researchers (with young children in tow), they couldn't hope to do so alone.[3] So they enlisted the help of birdwatchers.

Garnett inserted an article in the March 1993 issue of the *Bird Observer*, headed 'Are you visiting Cape York? Please help us to find the Golden-shouldered Parrot.' There were two ways that birdwatchers could help, he explained. One was to watch for the parrots at waterholes, and Garnett promised he could give good advice on which waterholes to watch, as well as access to some that were otherwise inaccessible. The other was to count the birds that came to drink, even if there were no Golden-shouldered Parrots among them. This would provide data on the parrot's prevalence relative to other seed-eating species, and could establish a baseline against which subsequent population changes could be measured. 'We can guarantee that you will see plenty of interesting birds and can almost guarantee you will see Golden-shouldered Parrots', Garnett advised prospective birdwatching helpers, but 'we cannot also offer material support'.[4]

He got lots of responses. The Golden-shouldered Parrot's combination of beauty and rarity had already made it a magnet for birdwatchers and twitchers, drawing a steady stream of traffic up the Peninsula Development Road to Musgrave and beyond. Birders jumped at the opportunity to boost the conservationist cause at the same time as seeing a rare bird. Among the dozens of volunteers who helped with the fieldwork were birding notables such as Andrée Griffin, Andrew Ley, Fred van Gessell, John Young, and Eric Zillmann. They counted birds and banded them, searched for nests and monitored those they found, measured vegetation plots, and in numerous other ways contributed to the success of Crowley and Garnett's investigations.[5]

It wasn't just amateur birders who helped those investigations. Like other recent conservation projects, saving the Golden-shouldered Parrot was a massive collaborative effort involving a vast diversity of interest groups and interested people. Successive issues of Crowley and Garnett's newsletter, *Antbed*, chart the growing corpus of collaborators, such that by the end of their three-year study the list of people thanked extended over two A4-sized pages.[6] Heading the list was the Shephard family of Artemis, followed by numerous other Cape York Peninsula pastoralists who helped in various ways. Added to those were Indigenous contributors, scientists, government departments, non-government agencies, universities, and museums. Birdwatchers contributed to Golden-shouldered Parrot research-and-recovery efforts, but as just one cog in a huge apparatus.

The apparatus worked. Crowley and Garnett made rapid progress toward answering their initial research question on the causes of the parrot's decline. Early in their investigations, they realised that the 'invasion of the grassy flats by tea-trees' posed a major threat to the species' habitat, and therefore to its survival, and that fire—or its lack—might be a factor.[7] 'We think that the parrots have declined because there has been too little fire not too much', they reported in April 1994. However, while they knew that the timing and intensity of fires were important, they admitted that, at this stage, 'we don't know exactly what sort of fires will best clean up the country'.[8] By late 1995, they had gone a long way toward finding out. A combination of hot, cool, and medium-intensity fires, lit strategically by season, was necessary to maintain the parrots' grassland habitat and to minimise predation, especially by butcherbirds. Open country is the parrot's first requirement, and fire is the traditional way of maintaining it.[9]

In their initial investigations, conducted between 1992 and 1995, Garnett and Crowley recognised that Golden-shouldered Parrots had flourished under Aboriginal environmental stewardship that included frequent fires, often of high intensity, and that the species had suffered under the more restrained burning regimes instituted by cattle graziers. 'Overall a reduction in burning frequency underlies all other threats to the survival of the Parrot', they wrote in October 1995, making it clear that the reduction was a consequence of the shift from Indigenous to European land-management strategies.[10] At this stage of their research, they consulted with local Aboriginal people, but Indigenous groups and knowledge had no special prominence in their investigations or remedial recommendations. That changed in the 21st century.

The Golden-shouldered Parrot's current range impinges on the lands of at least five Aboriginal groups—the Olkola, Wakamin, Thaypan, Kunjen, and Kokoberrin peoples, who name the bird variously as Alwal, Arrmorral, Thaku, and Minpin. As traditional owners of the lands on which the parrot retains its strongest presence (around Artemis), the Olkola people have taken a particularly prominent role in recent recovery efforts. For the Olkola, Alwal (as they call the Golden-shouldered Parrot) is an important totem whose survival is crucial to the health of country. More than that, they see the parrot's survival as bound up with the maintenance of their own cultural integrity and the proper custodianship of their lands. As expressed in a 2022 recovery plan to which the Olkola were leading contributors, recovery 'is not just about saving an endangered parrot but very much about ensuring people are on Country to look after Country properly'. Number one of the plan's nine objectives is to 'Return people to country to make golden-shouldered parrot habitat strong again.'[11] It's a conception of conservation that positions people not as problems to be overcome, but as assets to be encouraged.

The 2022 *Golden-shouldered Parrot Recovery Plan* proclaims itself 'the first Aboriginal-led recovery program in Australia'. While Aboriginal-led, the program takes 'a collaborative approach that incorporates Traditional Knowledge and western science', welding the two into an innovative and ambitious proposal to preserve the parrot. Or, as its executive summary puts it: 'This recovery plan is a two-way science document. The recovery actions have been developed through scientific knowledge and research, embedded within the cultural knowledge and ecological expertise of

Traditional Owners.'[12] Throughout the recovery plan, the imperative of collaboration is continually stressed, especially collaboration between scientists and traditional owners, but also partnerships with local pastoralists such as the Shephards of Artemis, government agencies such as the Queensland Parks and Wildlife Service, and non-government organisations such as Bush Heritage Australia.

Today's emphasis on collaboration in saving the Golden-shouldered Parrot contrasts with the situation of the Paradise Parrot 100 years ago. Even the points of commonality also highlight the differences. One point in common is the support of local graziers, the Shephards' position today being analogous with that of Cyril Jerrard in the 1920s. But Jerrard had only one supportive collaborator—Alec Chisholm, a journalist who had only his pen to promote the parrot's prospects. No one at the time had expertise in saving endangered species. The Shephards, by contrast, are the on-the-spot agents for a vast, multi-dimensional recovery effort, with enormous amounts of ecological expertise, Indigenous knowledge, scientific know-how, birders' observations, administrative acumen, and volunteer person-hours being funnelled through their station, Artemis. Sue and Tom Shephard know the Golden-shouldered Parrot intimately, but they're just the front line in the battle to save the species.

Large-scale collaborations linking a diversity of organisations and individuals are the norm in modern conservation campaigns. This the case for all Australian birds on the endangered list, not just the Golden-shouldered Parrot, but also the Regent Honeyeater, Plains Wanderer, Swift Parrot, Orange-bellied Parrot, and many more. Money and resources may be often inadequate; some iconic species may draw disproportionate attention; and partnerships may falter or fail. The point remains that it is now generally accepted that, as BirdLife Australia declares in promoting its 2023 'mission to save birds': 'When it comes to conservation, collaboration is key—we can't tackle the extinction crisis alone.'[13] Collaborative action has become a conservationist axiom.

Collaborations for conservation are not entirely new. They were conducted in earlier times, albeit on a modest scale. Campaigns against the plume trade in the early 1900s, for example, drew together interest groups ranging from ornithologists to feminists. From their inception in 1909, Gould Leagues were effectively conservationist collaborations between birding organisations and state education departments. The expansion of Australia's national parks in the mid-20th century was pushed by coalitions of birders, bushwalkers, naturalists, and adventurers. What's new about recent conservationist collaborations is their magnitude and their scientific framing.

The advent of large-scale conservationist collaborations was part of a thoroughgoing transformation and expansion of conservation from the 1960s onward. This was a global rather than merely an Australian phenomenon. Thomas Dunlap, among other historians, has taken the publication of Rachel Carson's *Silent Spring* in 1962 as a marker of the emergence of the environmental movement that brought a new sense of urgency to the preservation of life on planet Earth. Ecology was pushed to the fore, and its professional practitioners were elevated into guardians of the planet's fragile biodiversity. 'Biodiversity' itself was an ecological neologism, coined in the 1980s, joining other ecological buzzwords such as 'biosphere' (coined by ecologist G. Evelyn Hutchinson in 1970) as words in everyday usage. The aesthetic, emotional, and ethical imperatives of conservation were not lost, but ecology was given far greater weight than hitherto, and scientific expertise awarded an elevated importance.

It was not that scientists had the domain of conservation to themselves. Far from it. As conservation became more and more mainstream in the final third of the 20th century, it splintered along multiple lines, its exponents ranging from hippies to corporate executives. Nonetheless, conservation in the environmental age came increasingly under the aegis of scientists, and it became widely accepted that biologists in general, and ecologists in particular, were rightfully the arbiters of how nature could and should be preserved.[14] This was part of a wider societal process of professionalisation and elevation of the expert that gathered pace as the 20th century progressed.

Inevitably, this meant demotion for amateur birders. In earlier times, amateurs such as Arthur Mattingley were among the leaders of bird-conservation campaigns, with few scientists among their collaborators. By the latter decades of the 20th century, the progress of professionalisation had radically transformed the situation. Whereas they once had led conservationist actions, amateur birders now played second fiddle to scientists. Importantly, however, they weren't pushed out of the band. Bird conservation, like bird studies more generally, still welcomed the participation of amateur, unqualified bird observers, albeit under scientific guidance in projects that came increasingly to be called 'citizen science'.

For the birds, professionalisation was good news. Ecological experts acquired ever more sophisticated understandings of the factors that imperilled birdlife, and devised ever more sophisticated strategies for countering those perils. To put it simply, the more ecologically informed conservation became, the more potentially effective were the strategies proffered. Again, the histories of the Paradise and Golden-shouldered Parrots offer instructive comparison.

Cyril Jerrard and Alec Chisholm—both amateur birders—did what they could to save the Paradise Parrot in the 1920s. But they were stumbling in the dark. Their understanding of the causes of the parrot's decline, while apt enough at a high level of generality, lacked specificity. They knew it had something to do with cattle-grazing, fire, and predators, but had no understanding of how those factors interacted to bring the parrot down. On practical measures to reverse the decline, their suggestions were rudimentary and ineffectual, and no one else at the time had anything better to offer. By the 1990s, when professional scientists began detailed studies of the Golden-shouldered Parrot, things had changed dramatically. Conservation had undergone its seismic shift into ecology, so Crowley and Garnett came equipped with a vast array of conceptual tools, theoretical templates, research methodologies, prior knowledge, and advanced technologies unknown to Jerrard and Chisholm 70 years earlier.

Admittedly, there are still gaps and imperfections in scientific understandings of how to save our avifauna. And things didn't go smoothly for the Golden-shouldered Parrot. At the beginning of the 21st century, Garnett and Crowley optimistically suggested that its numbers and distribution may have been stabilising.[15] They weren't. Both numbers and range kept contracting, such that by 2021 only an estimated 50 remained at Artemis, long regarded as the stronghold of the species. So the Golden-shouldered Parrot Recovery Team redoubled their efforts and took ever more drastic action, even employing chainsaws and herbicides, as well as fire, to keep open the savanna habitat essential to the species' survival.[16] Such measures—treating trees as a threat to birdlife and promoting tree-killing as a conservationist tactic—may seem counter-intuitive.[17] But the fact that leading ecologists implemented these strategies testifies to their flexibility and their readiness to explore unconventional avenues for averting extinctions. Equally, the fact that pastoral lessees and traditional owners welcomed the adoption of such drastic tactics testifies to the strength of the collaborations that have been forged among these groups. There are solid grounds for hope.

There are also grounds for despair. The fate of the Golden-shouldered Parrot hangs in the balance. Today we're immensely better equipped to rescue it and other endangered species than was the case for the Paradise Parrot last century. But some impediments have roots deeper than science can reach. Despite the superior conservation strategies and technologies now available, the underlying drivers of extinction identified by Cyril Jerrard in 1924—'our avarice and thoughtlessness'—remain stubbornly persistent. As a society, we still too often prioritise economic gain over avian loss.

If we're to ensure that the Golden-shouldered Parrot and other endangered species do not go the way of the Paradise Parrot, we need ecological knowledge, and scientific strategies, and up-to-the-minute technologies. But we need more than those. Sometimes, at least, we need to subordinate avarice to avian welfare, and to do that we need to connect, emotionally and ethically, with the birds around us. Birds must matter to us, not merely in an abstract or objectified fashion, but as sentient beings of intrinsic worth. It's in fostering such an outlook that birdwatching can make one of its greatest contributions to conservation.

Of course, birdwatchers can and do contribute to conservation through participation in citizen-science projects. They conduct surveys and submit observations, and in myriad other ways add to the store of data on which effective conservation depends. However, birdwatchers can help endangered species such as the Golden-shouldered Parrot not only by such instrumental means, but also by strengthening the emotional bonds between people and birds. Birdwatching helps us reach across the species gap and connect with the wild in ways that are simultaneously emotional and empirical, sensory and cerebral, thereby deepening our appreciation of nature in manifold ways.

Conservation is, and always has been, as much an affair of the heart as of the mind. We can understand, intellectually, the importance of biodiversity and life's dependence on ecological interrelationships. For some people, such abstract understandings may be sufficient to sustain a commitment to conservation. But most of us need the reinforcement of experience: of touching the wild and feeling the thrill it brings; of encountering nature at first hand and knowing the elation that follows. Birdwatching is just one way of experiencing nature, but it's a way well suited to the exigencies of the modern world. It's a way that inculcates a conservation consciousness by fostering a love of birds, as those who promoted the pastime have long known. John Leach knew it when he launched his pioneering field guide in 1911, and his many successors have continued to encourage loving birds as a pathway to their preservation.

Loving birds won't tell us how to save those, such as the Golden-shouldered Parrot, that flutter on the edge of extinction. We need science for that. But it will tell us that they are worth saving.

CHAPTER TWENTY

Regent Bowerbird: Holidays

A REGENT BOWERBIRD'S RESPLENDENCE can take your breath away. Adult males are velvet black and vivid gold, crowned with a patch of brilliant flame-orange. Flying through the rainforest, they look like gigantic black-and-gold butterflies, their colours alternating as they flap and swerve. In pools of sunlight, they glow.

There's no better place to see Regent Bowerbirds than O'Reilly's Rainforest Retreat atop the Lamington Tableland in southern Queensland. And you needn't stop at seeing. At O'Reilly's, the birds have been fed for generations, so they'll obligingly hop onto your hand to give not only the best possible view, but also that tingle of excitement that comes from nature's touch. Their friendliness is a boon to photographers, too, although when John Bransbury was compiling *Where to Find Birds in Australia*—one of the first birding tourist guides to this country, published in 1987—he found O'Reilly's Regent Bowerbirds 'difficult to capture on film ... because in their eagerness to see what you have to offer by way of food they often approach too near to allow accurate focussing'.[1]

O'Reilly's Guesthouse was founded for nature-based tourism in 1926. By the time Peter O'Reilly began Bird Week 52 years later, the resort was revered as a birdwatcher's paradise. It was among the best places on Earth to see Albert's Lyrebirds, and one of a very few in Queensland that offered Rufous Scrub-birds and Olive Whistlers. There were also myriad less rare but no less spectacular birds, among them the Regent Bowerbird whose photograph adorned the flyer advertising the first Bird Week. Since then, the black-and-gold stunner has become O'Reilly's logo.

O'Reilly's first Bird Week in 1978 marked a milestone in the evolution of birdwatching holidays in Australia. Guests were almost guaranteed to see a suite of extraordinary birds under the guidance of six experts who included Roy Wheeler, Peter Slater, and Steve Parish. They not only guided guests to the birds, but also gave informative talks enlivened with state-of-the-art audio-visual presentations. Birders could socialise with like-minded enthusiasts and have fun in the outdoors, with comfortable beds and hearty meals awaiting indoors when night fell or the savage storms for which the district is infamous swept through. From the outset, Peter O'Reilly wanted Bird Week to be a 'light-hearted and enjoyable' event that deepened people's appreciation of birds in particular and nature more generally.[2] It was also a successful business venture.

Hosting birding holidays as a commercial enterprise is a relatively recent innovation, originating in the period of prosperity that began shortly after the end of the Second World War. Internationally, a landmark event was the foundation of Ornitholidays by Lawrence Holloway, who in 1965 took his first contingent of British birders to the Camargue in France.[3] Australians at that time had access to a limited range of birding-holiday options, but not to dedicated birding tourism operators such as Ornitholidays. Commercialisation would come to Australian birding holidays, though more slowly than in the UK and the US.

Non-commercial birdwatching holidaying has a much longer history. Just after the turn of the 20th century, the (R)AOU began a tradition of holding campouts immediately after its annual congresses. Usually two weeks in duration, campouts were occasions for birders to get together and have fun. Members brought their spouses and children, and organisers went out of their way to make the events family-friendly as well as ornithologically interesting.

The main excursion after the 1908 AOU congress was a boat trip through the Bass Strait islands that was considered too arduous and dangerous for family enjoyment. So the union put on an additional event, a campout on Phillip Island whose leader, A.J. Campbell, reported that the time 'was pleasantly spent egging, fishing, photographing, and observing'.[4] This event was attended by a mixture of men and women, girls and boys, whereas the Bass Strait boat trip had a complement of 25 men and just one woman, the redoubtable Mrs Ethel White. Gender disparities aside, both events were reported in the *Emu* in holiday spirit— the island campout as an occasion for family fun, and the marine excursion as a voyage of adventure, complete with mishaps and storms at sea.[5]

Even the RAOU congresses that preceded the campouts often included holiday elements. Between ornithological presentations at the 1924 congress in Rockhampton, attendees were encouraged to venture out and experience the local landmarks and birdlife at first hand. The program included excursions to the Botanic Gardens, Olsen's Caves, Fitzroy Vale, the Archer's Gracemere property, and Fairy Bower (a patch of vine-scrub nearby); there was also an outing up the river with the Fitzroy Motor Boat Club. All this was before the campout, which that year was at Byfield. On the way there, participants stopped at Yeppoon, where they went to the movies, took 'a dip in the sea', and had 'rambles round the township'. There were many more birding rambles around Byfield, interspersed with entertainments, including a lantern-slide night. After Byfield, many campout participants added an excursion to North Keppel Island.[6]

The holiday atmosphere of the Byfield event typified RAOU campouts in the interwar years. Birding, the core activity, was conducted in ways that maximised socialising and having fun. However, RAOU campouts were among the very few birding-holiday options then available, and most members' access to them was restricted by limited leisure time and a lack of reliable transport. The debacle of the Great Depression added to the difficulties.

Worse was to come. In the emergency of the Second World War, the RAOU held no congresses or campouts between 1940 and 1946. Determined birders still went on long-distance excursions, but wartime imposed exceptional impediments. Arnold McGill recalled a birding trip to Moree that he undertook with Jack Ramsay, Norman Chaffer, and Roy Cooper in late 1943. It was a 'time of severe petrol rationing':

> [So] it was necessary to tow a charcoal burner behind the car to generate the needed gas to drive the motor all the way. That in itself was a cumbersome added trailer, but worse, all the large car boot was used to store charcoal. All camping gear, photographic equipment, food, clothes etc. were packed to the roof on the back seat leaving just sufficient space to squeeze in one of the party so that three could ride on the front seat. Roy and I, as the two younger members, wedged ourselves in the back in turn.[7]

Nonetheless, McGill recalled it as 'a wonderful trip', the highlight of which was securing movie film of Spotted Bowerbirds.

After the war, campouts were not only revived but revivified into even more festive occasions than their pre-war predecessors. Birders, like everyone else, craved respite from the trauma of war. The second post-war campout, in September 1948, was especially notable for its high-spirited hijinks. Participants held mock quiz nights with silly and obscure ornithological questions; they wrote doggerel verses about their birding (and other) adventures, and sang them around the campfire. They even concocted a facetious magazine, the *Emulet*, lampooning the *Emu* and their own ornithological hobby. One item, 'The Epistle of Dom to the Philistines', ridiculed the divide between scientific ornithologists and recreational birdwatchers. Another

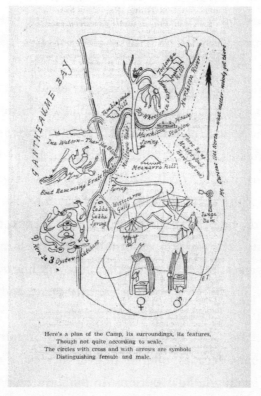

Figure 20. Eric Thake's zany map of the 1948 campout shows RAOU members letting their hair down on holidays. *Emulet*, 1949.

asked 12 absurd and unanswerable questions, the fourth of which was 'Who killed Cock Robin? Produce collector's license'.[8] It was a thinly veiled reference to the ugly event at the 1935 campout at Marlo, where licensed collector George Mack caused consternation by shooting a nesting Scarlet Robin in front of the campers. That the campers 13 years later were reducing the incident to farce suggests that they had put it behind them—and wanted to make a point of having put it behind them. Campouts were for fun, not bloodshed and controversy.

The 1951 campout at the Kulkyne State Forest in north-western Victoria resulted in a second *Emulet*, similarly filled with ornithological absurdities and holiday follies. No further *Emulets* were published, but campouts continued to be conducted in holiday mode, and many were held in touristy places such as Lake Barrine in 1953, Port Arthur in 1955, and Noosa Heads in 1958.

By the late 1960s, however, RAOU campouts were becoming more serious, research-orientated ornithological outings. Shedding some of their holiday qualities, they became shorter and attracted fewer participants. Libby Robin comments of this period that the 'expansion of private travel made large organised groups a less attractive option for ornithological holidaying'.[9] That may have been a factor, but so, too, was the RAOU's contemporaneous identity crisis as it traumatically transformed itself into a strictly scientific organisation. Australia's other major birding body, the Bird Observers' Club, suffered no such identity crisis, and its birding-holiday program went from strength to strength.

In 1946, the BOC established a Tours Committee, which, over subsequent decades, offered an ever-increasing number and diversity of birding-holiday options. In 1969, the club surveyed its members on their holiday preferences, asking whether they would prefer 'guesthouse-based' or 'safari' tours.[10] A few months later, it advertised its next holiday outing: a fortnight at Iron Range that was very much of the 'safari' kind. 'This will be a rough and tough camp in an isolated area and is not for the tourist but arranged for the dyed-in-the-wool birdwatcher', it warned, adding the big attraction of 'at least 20 birds found nowhere else in Australia'.[11] Despite that lure, such arduous holidays were not to everyone's taste, so the club also offered the option of ten days birding from the comforts of O'Reilly's Guesthouse.[12]

Into the 1970s and beyond, the *Bird Observer* featured more and more reports on BOC birding tours to places near and far, alongside announcements of forthcoming tours. As well as organised tours, birders also undertook increasing numbers of independent birding holidays, and reports on those, too, burgeoned in the pages of the *Bird Observer*. So did advertisements for accommodation to cater for those holidaymakers. From the 1970s, too, there was an upsurge in reports on, and advertisements for, birding trips overseas, including tours to such exotic places as the Silk Route in what was then Soviet Central Asia.[13]

Australia's boom in birding tourism in the second half of the 20th century followed Western-world-wide trends. Key factors were growing affluence, increasing leisure time, greater access to motor vehicles, including four-wheel-drive models, and the enormous expansion of air travel. Those same factors led to a massive growth of tourism of all kinds, and the boom in birding tourism was just one small part of that wider phenomenon.

These developments commercialised the birding holiday, enabling people to purchase their encounters with birds. Some might consider this a negative development, detracting from birding's aspiration to connect people with nature, and perhaps there's a measure of truth in that assessment. Yet, through a wider-angled lens, the growth of birding tourism can be seen as an outcome of the social transformations that resulted in more urbanised populations seeking recreation in less urbanised places. Ecotourism—of which birding tourism is a specialist type—commercialises the experience of nature, but touching the wild remains the compelling attraction.

Commercialisation notwithstanding, many birders still did their holidaying on the cheap. In a solo trip to Iron Range in May 1986, David Andrew flew from Cairns to Lockhart River, then 'hitched a ride 15 km. into the forest to an old tin shack that was to be my hotel for the next three nights'. The discomforts matched the price. His first night at Iron Range 'was among the most uncomfortable I have spent anywhere', he lamented, as he lay 'dripping with sweat, listening to marauding pigs and mosquitoes' while bush rats nibbled his toes and scrub-fowls screamed away any prospect of sleep. An abundance of birdlife, including several species new to him, was the reward, but Andrew sensed something was missing. With remarkable luck, he spotted a Red-bellied Pitta, and 'watched it, spellbound, until it bounced out of sight in the gloom of the forest. One of the unfortunate things about travelling alone was not being able to share this beautiful bird with anyone'.[14]

Andrew's words attest to the social dimension of birdwatching holidays. Some obsessive twitchers may choose to birdwatch alone, but the vast majority of birders prefer to share their experiences—disappointments as well as joys—with others, whether family, friends, fellow club members, or total strangers brought together for the event. Birding in company has practical benefits: it keeps down costs, and more eyes see more birds. But being with people who share the same passion is a stronger incentive to holiday together. It's clear from the numerous accounts of birding holidays in the *Bird Observer* and elsewhere that the writers revelled in the camaraderie of the campout and the conviviality of the coach tour.

The advent of more comfortable options, such as O'Reilly's Bird Week, probably intensified the social incentive.

With the late-20th-century surge in birding tourism came a new kind of guidebook. Australia's first finding guides were Roy Wheeler's state guides to New South Wales and Victoria, published in 1974 and 1979 respectively, followed by Michael Morcombe's *Great Australian Birdfinder* in 1986. John Bransbury's *Where to Find Birds in Australia* (1987) raised the Australian finding guide to new levels of sophistication and specificity, homing in on what to see where in a tersely informative style already familiar to visitors from the UK and US. For their convenience, he grouped sites into what he called 'holiday units, providing the visitor with a wide range of habitats and birds', thus maximising the bird count at the end of the trip.[15]

After Bransbury's came many more finding guides. Some were guides to specific regions, such as Jo Wieneke's *Where to Find Birds in North-east Queensland* (1992) and David Donato's *Finding Birds in Australia's Northern Territory* (1997). Some combined field guide with finding guide, as did Lloyd Nielsen's *Birds of Queensland's Wet Tropics and Great Barrier Reef* (1996). Some, such as Nigel Wheatley's *Where to Watch Birds in Australasia and Oceania* (1998), were explicitly aimed at international visitors. They were of varying quality, but one finding guide of this generation attained classic status: Richard and Sarah Thomas's *Complete Guide to Finding the Birds of Australia* (1996), affectionately known by birders as 'Thomas & Thomas'. This took the specificity of advice to new heights, telling birders exactly how far to travel down a particular track and which clump of trees to scrutinise to find the elusive species of their desire.

Finding guides promoted a twitchy style of birding. They're also used for casual birdwatching, but getting more ticks on the bird list is a prime purpose of a finding guide. Their authors have been up-front about this, none more so than Richard and Sarah Thomas, both of whom were avowed twitchers. A twitching orientation was evident in the first edition of their guide, but even more overt in the second, updated in 2011 by David Andrew and Alan McBride. In its new incarnation, it celebrated 'the new Holy Grail' of seeing 700 Australian species, and offered hope that readers might better it. It even forecast possible armchair ticks, advising, for example, that birders pay special attention to the Spectacled Monarch, as 'there is a potential split looming in the far north of Queensland'.[16]

Finding guides, birding tourism, and twitching are mutually supportive. Admittedly, birding tourism had a twitchy flavour even before Australia had finding guides. After all, seeing new species had always been an aim—or hope—of the birding holiday. Nonetheless, after the 1970s there was a noticeably twitchy trend in reports on birding tours in magazines such as the *Bird Observer*, whereby ever-increasing prominence was given to finding rarities, boosting lists, and enumerating 'lifers' (birders' jargon for a species not previously seen). To give just one example, Darrell Price's account of his 1988 trip to the Cape York Wilderness Lodge was arranged as a day-by-day list of his own and his companions' lifers, with running totals of both lifers and all birds seen.[17] By this time, such enumerations were commonplace in popular birding publications. A few decades earlier, they were virtually unknown.

Yet while the twitching element in birding tourism became more and more pronounced, a lot of holiday birding continued to be conducted casually. Shortly after Darrell Price published his piece in the *Bird Observer*, Susan Bailey of Ringwood published another in a quite different vein. Hers was a family holiday in which birdwatching was just one among many activities: an opportunity to share pleasant experiences of nature with her husband and daughters. Bailey's piece bears out the observation of Connie and John Trotter, earlier contributors to the magazine, who rounded off their stories about the incidental birding they undertook while travelling the Queensland coast by reflecting on 'how greatly a little bird-watching enhances a holiday anywhere'.[18] It's a reminder that alongside the escalation of twitching and commercialisation ran other trends that kept a great deal of holiday birding simple and relaxed. Just as AOU members in 1908 had the choice of the easy holiday on Phillip Island or the arduous one in Bass Strait, birders in more recent times could choose the level of birding intensity in their holidays.

Over time, the choices became wider and wider, with birding holidays available at every imaginable level of comfort and intensity (and price). Yet whether it's a full-on twitching odyssey to see every grasswren species in Australia, or a hosted birding and gastronomic getaway in a luxury villa, or a self-driven camping trip to see whatever comes along, the birding holiday is a way of playing in nature. And if nature's playground is exceptionally attractive, that's an added boon. Peter O'Reilly played that drawcard when he launched Bird Week in the scenic magnificence of the Lamington Tableland.

He knew, too, that birders came to the event to enjoy the company of other birders as well as of the birds. Bird Weeks featured campfire singalongs, poetry recitals, convivial competitions, and other entertainments designed to get people interacting. Reflecting on what made the event so special, Peter O'Reilly observed that the 'greeting of old friends and the making of new friends are all part of it, and birding is such a great people mixer'.[19] Bird Week, he explained, 'developed into something of almost religious significance where many people would return each year not so much to see new birds but to share the beauty of the rainforest and wildlife with their friends'.[20]

Prominent among the beauties in O'Reilly's rainforests are multitudes of Regent Bowerbirds. I'm sure most Bird Week participants have seen the species many times before, though perhaps not in such profusion. Yet the assembled birders still thrill at the sight. Partly, it's because they can get so close to such resplendent wild birds—close enough to touch them. But it's also because they can share the experience with others equally passionate about birds. Birdwatching is a way of connecting with nature. It's also a way of connecting with people, and the interplay between those two kinds of connecting has driven birding holidays since they began.

CHAPTER TWENTY-ONE

Superb Fairy-wren:
The Wild Near Home

SUPERB FAIRY-WRENS ABOUNDED IN the garden of the apartment in inner-suburban Canberra where I lived while researching this book. Iridescent-blue and purple-black males bounced across the lawn and scuttled into the shrubbery alongside their fawn-feathered female and juvenile companions. Superb Fairy-wrens are common in other cities, too, including Sydney, Melbourne, and Hobart, as well as in the farmlands and bushlands between those south-eastern capitals. In and around Perth, their place is taken by the even more dazzling Splendid Fairy-wren, the males sporting an electric violet-blue plumage that shimmers in the sunlight. In northern cities such as Darwin and Townsville, the common fairy-wren is the Red-backed species, coloured jet-black overall with a vivid scarlet saddle. There are six other species of fairy-wren scattered across Australia, ensuring that almost every town and city on the continent either hosts these gorgeous jewels of birdlife or at least has them living nearby.

Fairy-wrens are some of the most exquisite birds that can easily be found close to home, but they certainly aren't the only ones. Urban and suburban areas abound with birdlife. Where I live on Queensland's Sunshine Coast, we're visited daily by Pale-headed Rosellas, Rainbow Lorikeets, Yellow-tailed Black Cockatoos, Grey Butcherbirds, Lewin's Honeyeaters, Crested Pigeons, and many more. Occasionally, Rose-crowned Fruit-Doves and Regent Bowerbirds drop by; and there are several species, including Noisy Pitta and Russet-tailed Thrush, that we can hear in nearby bushland but never see. In eight years' residence here, I've recorded 94 species from my backyard. There's a local fairy-wren, too. Here, it's the Variegated Fairy-wren, rather like the Superb, but with bright chestnut shoulder patches.

With such a profusion of birdlife, it's unsurprising that lots of birding is done within urban bounds. In 1998, Birds Australia began a Birds in Backyards project to coordinate urban observations into a major research, education, and conservation program.[1] Since 2014, the same organisation, renamed BirdLife Australia, has conducted an annual Aussie Backyard Bird Count. It recently shed the second word, but most observations are still done in backyards.[2] Drawing tens of thousands of eager participants, according to BirdLife it's 'one of Australia's biggest citizen science events!' Its name is new, as is the electronic wizardry that gets data from suburban gardens into scientific datasets. But birding in backyards, like citizen science itself, is far from novel.

One hundred years ago, Harry Wolstenholme, son of the suffragette Maybanke Anderson, was an avid birdwatcher who did most of his watching in his garden in the northern Sydney suburb of Wahroonga. Sometimes, he backyard-birded alone; sometimes in company with birding legends of the day such as Keith Hindwood, Alec Chisholm, and Norman Chaffer. They not only admired

Wahroonga's birdlife; they meticulously recorded it and published their observations in the *Emu*. A glance through early issues of that journal reveals numerous articles on urban birds. One, by Wolstenholme in 1922, was a bird list for his suburb, with annotations combining affectionate appreciations with astute observations on each species. Superb Fairy-wrens (which he called Blue Wren-Warblers) he found especially charming, delighting in the 'bright warblings of these lovely little birds' that could 'be heard in every garden as they hop and flit about among the small plants and creepers'.[3]

Wolstenholme's own garden was an avian haven, arranged to encourage the birds to interact with him. To promote that process, he fed them, and, like others at the time, he had no compunctions about acknowledging the fact. Writing in the *Emu* in 1929, he explained how he fostered friendship with Superb Fairy-wrens: 'These little fellows, like many of the garden birds, are very fond

Figure 21. Harry Wolstenholme in his Wahroonga garden, feeding a Grey Shrike-thrush from his hand, c. 1928.

of cheese. While writing these notes on the verandah I have had to stop now and then to throw morsels to a pair of birds that came close below me in expectation of getting some.'[4] Wolstenholme not only fed his avian friends; he encouraged them to perch on his fingers as they did so. Quite a few obliged. His 1929 *Emu* article included a photograph of a Grey Shrike-thrush eating from his hand. He even fed a Lewin's Honeyeater by holding sugared water in his cupped palm while the bird perched on his fingers to lap up the sweet liquid. This was hands-on birding.

Recounting Wolstenholme's suburban birding exploits in his 1932 book *Nature Fantasy in Australia*, Alec Chisholm lauded such interactions unreservedly. Like Wolstenholme, Chisholm considered it wonderful that birds and people had built relationships of love and trust. He considered it wonderful, too, that such connections with wild birds could so readily be made in suburbia.[5]

Despite its title, *Nature Fantasy in Australia* covered only a small portion of the continent: the area within a 50-mile radius of Sydney's GPO. In it, Chisholm wrote rapturously about the birds, animals, and plants that dwelt there, and meditated on how the natural environment still shaped human life in and around Australia's biggest city. Setting the tone, the book's frontispiece is a painting by Neville Cayley captioned 'The Spirit of Sydney: Scarlet Honeyeater at nest in suburban garden.' The fact that this exquisite little bird was common in Sydney's gardens exemplifies Chisholm's theme of urban Australians' ready access to the wonders of nature.

That theme pervades all his writings, not just *Nature Fantasy*. One of Australia's most accomplished birders of the mid-20th century, Chisholm did most of his birding in and near the towns and cities of south-eastern Australia. He never visited the remote outback. Through his writings, he tried to persuade his compatriots to cherish the everyday birds, animals, and plants around them.

While sometimes he turned to more distant topics, he mostly celebrated the familiar nature that his fellow Australians could experience inside and just beyond their back fences.[6]

Chisholm was not alone in this. Many of his birding friends and contemporaries—including such notables as Keith Hindwood and Arnold McGill in Sydney, and Charles Bryant and Roy Wheeler in Melbourne—also wrote prolifically about the birds of urban and near-urban places, and, like him, did much of their birdwatching there. Partly, this was due to practicalities. Especially in the early decades of the 20th century, the difficulties of travel restricted birders' options, and put visiting remote places beyond the reach of many. Yet even as the outback opened up, as cars and roads improved, four-wheel-drive vehicles became available, and air travel became cheaper, birders continued to do much of their birding no more than an hour or two from home. The growing accessibility of elsewhere opened new options, but the familiar places near home—the 'local patch', to use the insiders' idiom—retained birdwatchers' loyalties.

From its inception in 1952 through to its demise over half a century later, the Bird Observers' Club's monthly magazine, the *Bird Observer*, devoted between a half and a third of each issue to birding in and near urban locations. Even as the magazine reported on birding excursions to ever more esoteric places in Australia and ever more exotic locales overseas, it continued to keep readers informed about the everyday birds its members encountered in their everyday lives.

Many *Bird Observer* articles reported observations from birders' backyards. Gay Grogan began one on birding in suburban Croydon in the June 1985 issue by exclaiming how 'absolutely delighted' she was 'to see the annual visit of the Fairy Wren to my backyard', then listed dozens of species she had seen there. Entranced by 'their graceful aerial ballet' and entertaining antics, she intimated that the 'gamut of emotions' stirred by the birds in her backyard made her life richer and more fulfilling.[7] Barbara Burns of Templestowe contributed a piece on 'Birds on a busy schedule', fondly describing the many birds she encountered during her daily routines of housekeeping, taking the kids to school, and heading off to work. By putting her in touch with the wild, birds offered respite from mundane matters, even though the wild with which she connected was just off a suburban street.[8]

In the 1990s, Molly Brown regularly contributed articles to the *Bird Observer* on the birds she saw around her home in Manjimup, Western Australia. In typical birdwatcher fashion, she combined expressions of affection for the birds with acute observations about their behaviour. An article on 'Our resident swallows' inquired into those birds' breeding and migratory habits, while another on a 'Roadside walk' discussed birds' adaptations to roads and roadside vegetation corridors. In 'Watching at the window', she delighted in the birds she saw through the casement windows of her living room, without venturing outside. 'A variety of birds come', she wrote, 'but star billing is given to the Splendid Fairy-wren and the Red-winged Fairy-wren. A full plumage male fairy-wren must be high on the list of the world's most beautiful birds, and ... fairy-wrens have endearing characters, too.'[9]

Urban birding extends far beyond admiring fairy-wrens from the comfort of the lounge-room. It can uncover some out-of-the-way birds in some not-so-pretty places. Arnold McGill, reminiscing in 1980 on half a century of birding in Sydney, recalled with fondness the Malabar headland 'where the sewer outfall attracted a great number of sea-birds', making it 'as good as any place in the world to watch sea-birds, outside the Antarctic'. He had seen as many as 548 Wandering Albatrosses there in a single day, as well as Black-browed and White-capped Albatrosses, Giant Petrels, and several species of shearwater and prion. Alas, 'the so-called march of progress' put an end to that superb site for seabird-watching in suburban Sydney.[10]

Melbourne birders were, and still are, able to get more up-close and personal with sewage-loving birds. Werribee Sewage Farm, now blandly—and therefore inaptly—renamed the Western Treatment Plant, lies just off the freeway between Melbourne and Geelong. So prolific is its birdlife that Werribee has been declared a Ramsar site, and attracts birders not only from Melbourne, but from all over Australia and beyond. Twitcher Sue Taylor maintains that 'it is impossible to have a bad day birding at Werribee'.[11] Waders and waterbirds are the most abundant attractions, but Werribee also hosts numerous other species, including the critically endangered Orange-bellied Parrot and the closely related Blue-winged Parrot. A sewage farm may not meet everyone's ideal as a place to commune with nature, but it provides crucial habitat for rare and threatened species, including some that come halfway around the world to banquet on its abundance.

Smaller though similarly smelly urban locales have long attracted birders as well as birds. In 1947, Charles Bryant published a tribute to the birds of Fishermen's Bend near Port Melbourne. Subtitling his article 'Beauty in a Municipal Garbage Tip', Bryant revelled in both the diversity and the tenacity of the birds to be found there. 'Undeterred by the mephitic aroma of the burning tip, the birds go about their lawful occasions', he observed; and they were just as beautiful and just as fascinating as birds in less easily accessible—if also less malodorous—places.[12] This, and the numerous other instances of birders delighting in rubbish dumps, sewage farms, and the like, testifies to the transformative magic of birds, enabling us to penetrate beyond the superficial unsightliness—and smelliness—of such places and there behold the mysteries of nature.

The romance of birds may be transformative, but some birders took the romance of the sewage plant literally. Graham Pizzey, recounting the lead-up to his marriage to Sue Taylor in 1957, remarked that, 'Quite often ... we did our courting at the Werribee sewerage farm.'[13] It's clear from his recollections that the choice of place was his, not Sue's. Still, as Pizzey went on to explain, his romance in a sewage farm led into a long and happy marriage, and there's something delightfully apt about one of Australia's greatest field guide authors fluffing his courtship feathers at a place whose richness in birdlife resulted from some of humanity's baser functions.

Sewage farms are the haunt of dedicated birders, but less smelly suburban sites offer equally engrossing birds. Legendary twitcher Sean Dooley began his zany birdwatching guidebook, *Anoraks to Zitting Cisticolas*, by describing an encounter with a flock of Musk Lorikeets in a busy carpark in the bayside Melbourne suburb of St Kilda. Although he loved seeing rare birds in far-flung places, Dooley admitted 'the truth is that these car park lorikeets a mere five minutes from my home offer the quintessential birding experience'.[14] It's not just that the lorikeets are pretty, although that helps. More importantly, they juxtapose the wild and the human in ways that illuminate our continuing connectedness with nature. The birds impress themselves upon our senses; they confront us with their raucously colourful reality; they tell us, visually and vocally, that even here, in a superficially soulless suburban carpark, nature not only survives, but thrives.

As Dooley attests, birding at a local patch remains popular even among twitchers. Convenience, undoubtedly, is a factor, but there's more to it than that. It's also because birders—including twitchers—treasure the birds near home. Birding near home may seldom secure new ticks for seasoned birdwatchers. Yet while most birders enjoy adding new ticks to their lists, very few are interested in nothing but ticking. For most birders, today as in the past, a core component of birding is encountering the wild; and the wild near home can be as fascinating, as puzzling, as beautiful, and as awesome as the wild further afield.

Moreover, engaging with the wild near home has an appeal of its own, since it attunes us to the rhythms and syncopations of nature that throb through even the urban environs where most of us live. Hearing the first Koel in September, or the Pied Butcherbird singing every day of the year; watching Superb Fairy-wrens transmute from drab to debonaire as breeding beckons, or Galahs clowning in unvarying costumes of pink and grey; being alert to the flitting, fleeting influx of Scarlet Honeyeaters when the right trees blossom, or the eternal, exasperating presence of Noisy Miners: these and hundreds of other common avian activities have captivated birdwatchers since the pastime's inception, and drawn them into an intimate awareness of how nature changes, remains constant, and alternates between the two. They're among the innumerable interactions with birds we can experience close to home, inducting us into a world beyond humanity, but not beyond our capacity for empathetic connection. Alertness to birdlife enhances our appreciation of home by making us aware of its proximity to the wild.

In some ways, the intimacy with nature that birders seek is easier to find when engaging with the familiar birds around home than with unfamiliar species in far-flung places. Birding close to home may not be a wilderness experience, but it's a way of touching the wild, with all the wonderment that can arouse. A Superb Fairy-wren in the backyard is as wondrous as a Purple-crowned Fairy-wren in a pandanus thicket beside a river in the Kimberleys, even if they're wondrous in somewhat different ways.

The Superb Fairy-wren in the backyard is an emissary of the wild. So is its Purple-crowned cousin beside that river in the Kimberleys, but it lives in a place we automatically recognise as wild, whereas the Superb dwells in the midst of human artifice. Its presence there reminds us of the resilience of the wild, its persistence even in environments humans have radically transformed. The resilience of the wild is not boundless, and if we're prepared to listen, the fairy-wren and its backyard bird companions might tell us of the need to protect their homes by preserving the remnants of nature around our own. Backyard birds are living proof that wild creatures can thrive in the interstices between the humanised and the natural worlds. But none can survive the obliteration of the wild.

Backyard birds are, mostly, common species; and commonness can dim appreciation, as I've earlier noted in relation to the beauty of Galahs. Yet the other side of commonness is familiarity, which can boost our appreciation of birds, fostering closer and more amiable relations. That's what Neville Cayley was intimating when he nominated the Superb Fairy-wren as Australia's 'favourite' among 'this family of feathered jewels':

> Perhaps this is because he is abundant in the more populous parts and therefore is more closely associated with our home life. He is quite common in gardens, both public and private, right in the heart of our cities and towns.... No matter where you meet him, you will always find him a charming, trustful little bird, ever ready, with a little encouragement, to make friends.[15]

There's more than a touch of anthropomorphism in Cayley's words, but that enhances, rather than diminishes, the point he's making. As classical scholar and birder Jeremy Mynott argues, 'some degree of anthropomorphism is probably both unavoidable and positively desirable' when we bring birds into our orbit of understanding and empathy.[16] That's especially the case for the birds who live daily among us.

Through birders' writings on backyard birds, stretching back more than a century, flows a message that, beneath the superficial ordinariness of suburbia, extraordinary things can be found if only we take the trouble to look. The same message is evident in bird art. Peter Trusler's paintings in the 1980 coffee-table book *Birds of Australian Gardens* are exemplary, conveying the beauty of suburban birds with stunning realism. Many are depicted alongside artificial props: the Willie Wagtail perches on a garden tap; the Red Wattlebird, on a plastic feeder; the Laughing Kookaburra, on a timber table with a lump of minced meat beside it; and a pair of Superb Fairy-wrens disport among exotic creepers in a rock garden. Making the message explicit, Trusler tells us in an 'Artist's Note' that he 'tried to capture something of the "living magic" that the authors and I find as we watch birds go about their daily activities. It can be just as equally appreciated in the man-made tapestry of the urban environs as in the natural splendour of the wilds'.[17]

As Trusler's words attest, there's a dichotomy built into our language that divides the urban from the wild. Yet one of the wonderful things about birding—and this is a point Trusler was getting at—is its revelation that birds bridge that divide. Birds live wild in urban spaces. They are, usually, the most visible, most beautiful, and most captivating wild creatures that we encounter around our homes. A compelling motive behind birdwatching is to touch the wild, and the fact that we can fulfil that primal desire in places near home adds to the pastime's appeal. Of course, birders also venture further afield; and as the capacity to do so has expanded, via cars, planes, and all the wonders of modernity, so birders' horizons have widened. But birders still take joy in fairy-wrens bouncing brightly around the backyard, magpies carolling from the clothesline, and sunbirds nesting on the back verandah. These are simultaneously ordinary and extraordinary: vignettes of an avian world so close we can touch it, so similar we can embrace it, yet so different from the business of humans that we can never fully comprehend it.

Figure 22. Superb Fairy-wren in a Canberra backyard.

Acknowledgements

HEARTFELT THANKS TO MY wife, Christine Mitchell, for encouraging, critiquing, and enduring my writing of this book. Thanks, too, to my daughter, Caitilin, and my son, Lachie, for making it all worthwhile.

Warm thanks to my agent, Marg Gee, for successfully parking my manuscript with a wonderful publisher, and to my publisher, Henry Rosenbloom, for skilfully steering it into press. Thanks, too, to the great team at Scribe who adroitly negotiated every bend and bump, getting to the destination well before the scheduled time.

Special thanks to Sarah Pizzey and Raoul Slater for helping boost this book into the celebration of birds it has become. Sarah not only gave open access to the papers of her father, Graham, but also provided copies of many of them and explanations of their contents. Raoul allowed me to include the glorious bird paintings by his father, Peter, as well as some of his own superb photographs. Both Sarah and Raoul offered personal insights into the lives and careers and two of Australia's greatest field guide authors.

For other exquisite artworks that enhance this book, I'm indebted to Wendy Cooper, Sally Elmer, Frank Knight, and Peter Trusler. I thank them all for their generosity.

Thanks, too, to the other institutions and people who allowed me to include their illustrations in this book: the Australian Museum Archives, BirdLife Australia, Allan Burbidge, the Gould League, Vince Lee, the Mitchell Library, the National Library of Australia, the Roger Tory Peterson Institute, the Royal Society for the Protection of Birds, and the Royal Zoological Society of New South Wales.

Other birders to whom I'm indebted are innumerable, but for their assistance and advice – or simply for making my birding excursions more wonderful than they otherwise would have been – I'd like to thank Shane Bennett, Walter Boles, Tess Brickhill, Allan Burbidge, Patricia Burke, Andrew Cockburn, Pat Comben, Ken Cross, Gabriel Crowley, Geoffrey Dabb, Alan Danks, George Diggles, Berry Doak, Reg Doak, Lesley Eagles, Cecile Espigole, Stephen Garnett, Denise Lawungkurr Goodfellow, Andrew Isles, Darryl Jones, Leo Joseph, John Kooistra, Claude Lacasse, Russ Lamb, Vince Lee, Timothy Nevard, Penny Olsen, Carol Popple, Steve Popple, Peter O'Reilly, Harry Recher, Greg Roberts, Carolyn Scott, Glen Threlfo, Peter Valentine, Brian Venables, and Phil Venables. Some of these people are professional ornithologists; some are recreational birders; but as this book shows, we shouldn't make too much of the distinction.

Much of the research for this book was done through a fellowship at the National Library of Australia. Thanks to the staff there, especially Simone Lark, Sharyn O'Brien and Andrew Sergeant. Some earlier research was conducted under a fellowship at the State Library of New South Wales, among whose staff I'd like to especially thank Richard Neville and Cathy Perkins. Thanks, too, to

Rachael Kosinski at the Roger Tory Peterson Institute in Jamestown, New York, who offered invaluable assistance in locating material on Peterson's Australian connections.

Thanks to historians Emily Gallagher, Tom Griffiths, and Libby Robin, and anthropologists Philip Clarke and Peter Sutton for the expertise they so freely offered.

Others to whom I owe a debt of appreciation include Larry Crook, Kylie Freebody, Robin Hill, Danielle Jesser, and Pat Noonan. Maybe I could have included them in the category of birder, but it somehow seems appropriate to separate them out. I'm not entirely sure why, because the line between birder and non-birder often blurs. I hope this book blurs it further.

List of illustrations

Figure 1 Sid Jackson (seated, centre) with Aboriginal assistants in their Cherra-chelbo camp, 1908. [National Library of Australia, PIC ALBUM 1243/3 #PIC P887/1360-1404]

Figure 2 Two of Jackson's Aboriginal assistants: on the left is the Yidinji man known to Jackson as 'Mitchell'; the man on the right may have been the one whom he called 'Billy'. [National Library of Australia, PIC ALBUM 1243/3 #PIC P887/1360-1404]

Figure 3 Australia's first bird photograph: A.J. Campbell's 'Rookery of Crested Terns by ocean', 1889. [National Library of Australia, PIC BOX PIC/7586 #PIC/7586/244]

Figure 4 'Taking a Wood-Duck's nest' by A.J. Campbell, 1894. [National Library of Australia, PIC BOX PIC/7586 #PIC/7586/7]

Figure 5 Sandwich-board men hired by the RSPB in London to carry the anti-plume-trade message via Mattingley's distressing photographs. [RSPB, rspb-images.com]

Figure 6 Chisholm's cumbersome photographic contraption. [Mitchell Library, PXA 1772, box 3]

Figure 7 Pioneers of the Australian field guide: John Leach (front); Neville W. Cayley (behind). [National Library of Australia, PIC BOX PIC/7586 #PIC/7586/176]

Figure 8 Monochrome illustration from J.A. Leach's *An Australian Bird Book*, 1911. [Courtesy of National Library of Australia]

Figure 9 The hessian hide from which Cyril Jerrard took the first—and only—photographs of Paradise Parrots. [Australian Museum Archives]

Figure 10 Male Paradise Parrot at nest-hole in a termite mound; photographed by Cyril Jerrard in March 1922. [National Library of Australia, PIC Box PIC/8902 #PIC/8902/3]

Figure 11 Female (top) and male (below) Paradise Parrots: one of only four or five photographs of this species ever taken, all by Cyril Jerrard. [National Library of Australia, PIC BOX PIC/8902 #PIC/8902/1]

Figure 12 Cyril Jerrard inspecting an abandoned Paradise Parrot nest in a termite mound, 1920s. [Mitchell Library, PXA 1772, box 6]

Figure 13 Norman Chaffer's Rock Warbler frontispiece for Hindwood and McGill's *The Birds of Sydney*, 1958. [Courtesy of Royal Zoological Society of New South Wales and National Library of Australia]

Notes

Preamble: Capricorn Yellow Chat

1   Peter Slater, *Rare and Vanishing Australian Birds*, Rigby, Adelaide, 1978, p. 86.
2   The following five paragraphs rely mainly on Allan Briggs, *The Story of the Capricorn Yellow Chat*, Allan Briggs, Rockhampton, 2016.
3   A.J. Campbell, 'The Yellow-breasted Bush-Chat (*Ephthianura crocea*)', *Emu*, 17: 2, 1917, p. 61.
4   Slater, *Rare and Vanishing*, p. 86.
5   Quoted in Briggs, *Yellow Chat*, p. 39.
6   Neville W. Cayley, *What Bird Is That?*, Angus & Robertson, Sydney, 1931, p. xvii.
7   Peter Slater, Sally Elmer, and Raoul Slater, *Glimpses of Australian Birdlife*, New Holland, London, 2018, p. 237.
8   Graham Pizzey, 'Foreword' to Roger Collier, John Hatch, Bill Matheson and Tony Russell (eds), *Birds, Birders & Birdwatching 1899–1999: Celebrating one hundred years of the South Australian Ornithological Association*, SAOA, Adelaide, 2000, p. vii.
9   https://exislepublishing.com/product/getting-closer-bird-photography/
10 Alec Chisholm, 'Our charming birds', *Telegraph*, 17 September 1926.
11 Stephen Moss, *A Bird in the Bush: A social history of birdwatching*, Aurum, London, 2004; Thomas Dunlap, *In the Field, Among the Feathered: A history of birders and their guides*, Oxford University Press, Oxford, 2011; Scott Weidensaul, *Of a Feather: A brief history of American birding*,

Harcourt, Orlando, 2007; Mark Barrow, *A Passion for Birds: American ornithology after Audubon*, Princeton University Press, Princeton, 1998.

Chapter 1: Galah

1   Joseph Forshaw, 'Galah', in H.J. Frith (ed.), *Birds in the Australian High Country*, Reed, Sydney, 1969, p. 224.

2   Bill Gammage, 'Galahs', *Australian Historical Studies*, 40: 3, 2009, pp. 275–93.

3   Editorial note on W.J. Tohl, 'Galahs come south: South Australian experiences', *Wild Life*, 3: 3, 1941, p. 107.

4   Stephen Debus, 'Review—*The Galah* by Ian Rowley', *Australian Bird Watcher*, 14: 1, 1991, p. 35.

5   Dom Serventy, 'Foreword' to Joseph Forshaw, *Australian Parrots*, Lansdowne, Melbourne, 1969, p. xi.

6   Graham Pizzey, 'Why not Kruger Parks here?' *Herald*, December 1965; clipping in Papers of Graham Pizzey, courtesy of Sarah Pizzey, Victoria.

7   Julian Huxley, *Bird-watching and Bird Behaviour*, Chatto & Windus, London, 1930, p. 26.

8   J.A. Leach, *An Australian Bird Book: A pocket book for field use*, Whitcombe & Tombs, Melbourne, 1911, pp. 93–94.

9   Charles Barrett, *From Range to Sea: A bird lover's ways*, Lothian, Melbourne, 1907, p. 34.

10   Neville W. Cayley, *Australian Parrots: Their habits in the field and aviary*, Angus & Robertson, Sydney, 1938, p. 245.

11   A.J. Campbell, 'A naturalist's camp in northern Queensland', no. 5, *Australasian*, 22 May 1886, p. 44.

12   Frank Littler, *A Handbook of the Birds of Tasmania and its Dependencies*, Frank Littler, Launceston, 1910, pp. 3–4.

13   A.J. Campbell, *Nests and Eggs of Australian Birds: Including the geographical distribution of the species and popular observations thereon*, A.J. Campbell, Sheffield, 1900, vol. 1, p. 5.

14   Slater, *Rare & Vanishing*, p. 26.

15   David Hollands, *Eagles, Hawks and Falcons of Australia*, Nelson, Melbourne, 1984, p. 73.

16   Michael Sharland, *A Territory of Birds*, Rigby, Adelaide, 1964, pp. 213–14.

17   A.H.S. Lucas and W.H. Dudley Le Souëf, *The Birds of Australia*, Whitcombe & Tombs, Melbourne, 1911, p. 346.

18   Russell McGregor, *Idling in Green Places: A life of Alec Chisholm*, Australian Scholarly Publishing, Melbourne, 2019.

19   Jennifer Ackerman, *The Bird Way: A new look at how birds talk, work, play, parent and think*, Scribe, Melbourne, 2020.

20 Lucas and Le Souëf, *Birds*, p. 197.
21 Gisela Kaplan, *Bird Minds: Cognition and behaviour of Australian native birds*, CSIRO Publishing, Melbourne, 2015, p. v.
22 Serventy, 'Foreword', p. x.

Chapter 2: White-throated Gerygone
1 Leach, *Bird Book*, p. 124.
2 Michael Morcombe, *Field Guide to Australian Birds*, Steve Parish Publishing, Brisbane, 2000, p. 240.
3 Arnold McGill, *Australian Warblers*, Bird Observers Club, Melbourne, 1970, p. 10.
4 'Recommended English names for Australian birds', *Emu*, 77: supp. 1, 1978, p. 303.
5 Rex Sharrock, '"New" names for old ones?', *Bird Observer*, 571, 1979, p. 35.
6 Rex Buckingham, '"A Birdo's Lament" or "More Names, Less Birds"', *Bird Observer*, 568, 1979, p. 12.
7 McGill, *Warblers*, p. 119.
8 Ibid., p. 125.
9 Dom Serventy and Hubert Whittell, *A Handbook of the Birds of Western Australia (with the exception of the Kimberley Division)*, Paterson's Press, Perth, 1948, p. 272.
10 Tim Low, *Where Song Began: Australian birds and how they changed the world*, Penguin, Melbourne, 2014, pp. 77–78.
11 Nigel Butterley to Alec Chisholm, 8 April 1965, Alec H. Chisholm Papers, Mitchell Library (hereafter ML), MSS 3540, box 5009; Alec Chisholm, 'Bird or wandering voice?' *Sydney Morning Herald*, 13 March 1965, Chisholm Papers, box 4983.
12 Russell McGregor, 'Mateship with nature: Nationalism and conservation in the writings of Alec Chisholm', *Environment and History*, 27: 3, 2021, pp. 399–420.
13 See for example Harry Wolstenholme, 'British and Australian birds—a comparison', *Emu*, 26: 2, 1926, pp. 131–33.
14 McGregor, *Idling*, p. 115.
15 Alec Chisholm, 'Do you know our elfin warblers?', *Wild Life*, 4: 1, 1942, p. 7.
16 Sidney Jackson to H.L. White, 26 September 1920, Papers of Sidney William Jackson, National Library of Australia (hereafter NLA), MS 466, box 5, item 166.
17 Michael Sharland, *Tasmanian Birds: How to identify them*, Oldham, Beddome & Meredith, Hobart, 1945, pp. iv–v.

18 Edward Sorenson, 'Australian bird clatter', *Clarence & Richmond Examiner*, 25 February 1908, p. 6.

19 E. Swan, 'What do birds say', *Australian Women's Mirror*, 2: 5, 29 December 1925, p. 27.

20 Robert Hall, 'Morning song of the Noisy Miner (Myzantha garrula)', *Emu*, 15: 3, 1916, pp. 185–87.

21 Weidensaul, *Of a Feather*, p. 211.

22 Cayley, *What Bird?*, p. 109.

23 Peter Slater, *A Field Guide to Australian Birds: Volume Two: Passerines*, Rigby, Adelaide, 1974, p. 141.

24 Peter Slater, Pat Slater and Raoul Slater, *The Slater Field Guide to Australian Birds*, Rigby, Sydney, 1986, p. 262

25 *The Graham Pizzey & Frank Knight Field Guide to the Birds of Australia*, HarperCollins, Sydney, 1997, p. 360.

26 Peter Menkhorst, Danny Rogers, Rohan Clarke, Jeff Davies, Peter Marsack and Kim Franklin, *The Australian Bird Guide*, CSIRO Publishing, Melbourne, rev. ed., 2019, p. 356.

Chapter 3: Tooth-billed Bowerbird

1 Sidney Jackson, 'In the Barron River valley, north Queensland', *Emu*, 8: 5, 1909, p. 236

2 Barrow, *Passion*; Moss, *Bird in the Bush*.

3 Leo Joseph, 'Museum collections in ornithology: Today's record of avian biodiversity for tomorrow's world', *Emu*, III: 3, 2011, pp. i–xii.

4 Sidney Jackson, Dorrigo Trip, 1910, 4 December, Jackson Papers, box 3, item 133; Sidney Jackson, 'The haunt of the Rufous Scrub-bird (Atrichornis rufescens, Ramsay)', *Emu*, 10: 5, 1911, p. 336.

5 Sidney Jackson, Narrative diary of a trip to Queensland, 9 June 1908–19 January 1909, Jackson Papers, box 3, items 126-130, p. 267.

6 The Yidinji are the traditional owners of the land on which Jackson camped and did most of his collecting. Some of the Aboriginal people with whom he interacted may have come from neighbouring groups such as the Ngadjon-ji.

7 Penny Olsen and Lynette Russell, *Australia's First Naturalists: Indigenous peoples' contribution to early zoology*, NLA Publishing, Canberra, 2019.

8 [Sidney Jackson,] *Egg Collecting and Bird Life of Australia: Catalogue and data of the "Jacksonian Oological Collection"*, F.W. White, Sydney, 1907, p. 1937.

9 Jackson, 'Barron River', p. 257.

10 Jackson, Diary, pp. 264, 269, 364, 366, 423.

11 Ibid., pp. 350–51, 356.

12   Jackson, 'Barron River, p. 272.
13   Jackson, Diary, pp. 357–59.
14   Ibid., p. 364.
15   Ibid., p. 389.
16   Ibid., pp. 414–16.
17   Moss, *Bird in the Bush*, p. 53.
18   Jackson, Diary, p. 130.
19   Ibid., p. 294.
20   Ibid., p. 352.
21   Ibid., pp. 376–79.
22   Jackson, *Egg Collecting*, p. vii.

Chapter 4: Crested Tern

1   Peter Slater, *Masterpieces of Australian Bird Photography*, Rigby, Adelaide, 1980, pp. 35–36.
2   Ian Campbell to Dom Serventy, 9 April 1981, Papers of Archibald J. Campbell, NLA MS 9650, box 1, folder 1.
3   Campbell, *Nests and Eggs*, vol. 1, p. 42; vol. 2, p. 838.
4   Campbell, *Nests and Eggs*, vol. 2, pp. 838–39.
5   Ian Abbott, 'John Gould's foremost successor in Australia: A.J. Campbell's travels in search of birds, their nests and eggs, and his contribution to ornithological knowledge', *Victorian Naturalist*, 138: 1, 2021, p. 17.
6   Campbell, *Nests and Eggs*, vol. 1, pp. 19, 67.
7   Ian Mason and Gilbert Pfitzner, *Passions in Ornithology: A century of Australian egg collectors*, CanPrint Communications, Canberra, 2020; Abbott, 'Gould's foremost successor'.
8   Campbell, *Nests and Eggs*, vol. 2, pp. 705, 882–92, 1064–65.
9   A.J. Campbell, 'Camp-out on Phillip Island', *Emu*, 8: 4, 1909, pp. 207–10.
10   Campbell, *Nests & Eggs*, vol. 2, pp. 762–64, 885.
11   Charles Belcher, *Birds of the District of Geelong, Australia*, W.J. Griffiths, Geelong, 1914, p. vii.
12   A.J. Campbell, 'A naturalist's camp in northern Queensland', part 8, *Australasian*, 4 September 1886, p. 39.
13   Jack Hyett, *A Bushman's Year*, Cheshire, Melbourne, 1959, for example pp. 35, 133–35.
14   Hugh Officer, *Recollections of a Birdwatcher*, Hawthorn, Melbourne, 1978, p. 106.
15   Ken Simpson and Zoë Wilson, *Birdwatching Australia & New Zealand*, Reed New Holland, Sydney, 1998, p. 175.

16 R.T. Littlejohns, 'Hunting birds with a camera', *Bird Lover*, 2, 1949, pp. 12–15.

17 Peter Slater, *The Birdwatcher's Notebook*, Weldon, Sydney, 1988, p. 51.

18 Donald Trounson, 'National Photographic Index of Australian Birds: Book Project', February 1967, pp. 4–5, Papers of Donald Trounson, NLA MS 9625, box 1.

19 See, for example, note by H[enry] K[endall] on Fred Berney, 'North Queensland notes on some migratory birds', *Emu*, 4: 2, 1904, p. 47.

20 Edmund Selous, *Bird Watching*, Dent, London, 1901, pp. 335–37.

21 Charles Barrett, *Koonwarra: A naturalist's adventures in Australia*, Oxford University Press, London, 1939, p. 99.

Chapter 5: Plumed Egret

1   'The Australasian Ornithologists' Union', *Emu*, 1: 1, 1901, p. 4.

2   Frank Littler, 'Bird Protection', *Emu*, 1: 1, 1901, p. 12.

3   Ibid., pp. 10–11.

4   Belcher, *Birds of Geelong*, pp. vii–viii.

5   Thomas Stephens, 'Need for bird protection', *Emu*, 14: 1, 1914, pp. 60–61.

6   Wild Life Preservation Society of Australia, *The Tragedy of the 'Osprey' Plume*, c.1910, Papers of Vincent Serventy, NLA MS 4655, box 197.

7   A.H.E. Mattingley, 'Plundered for their plumes', *Emu*, 7: 2, 1907, pp. 71–73.

8   *The Story of the Egret in Seven Scenes*, Royal Society for the Protection of Birds, London, 1909.

9   Weidensaul, *Of a Feather*, pp. 154–170; Dunlap, *In the Field*, pp. 30–34.

10 See, for example, Stephen Moss, *Birds Britannia: How the British fell in love with birds*, HarperCollins, London, 2011, pp. 84–87.

11 Susan Magarey, *Passions of the First Wave Feminists*, UNSW Press, Sydney, 2001.

12 Penny Paton, 'Murderous millinery: History of the bird plume trade with special reference to South Australia', *The Birder*, 253, 2020, pp. v–vi.

13 K[atherine] S[usannah] P[ritchard], 'Women and birds: Prohibition-of-Plumage Bill,' *Herald* (Melbourne), 13 November 1908, p. 6.

14 A.J. Campbell, 'Annotations', *Emu*, 7: 2, 1907, pp. 91–92.

15 'Bird Protection', *Emu*, 10: 2, 1910, pp. 142–43.

16 'Bird Protection in America', *Emu*, 1: 3, 1902, pp. 142–43

17 Stephens, 'Need', pp. 60–62.

18 Libby Robin, *The Flight of the Emu: A hundred years of Australian ornithology*, Melbourne University Press, Melbourne, 2001, p. 97.

19 Stephens, 'Need', p. 60.

20 Walter Froggatt, *Some Useful Australian Birds*, Government Printer, Sydney, 1921, p. 5.

21 S.A. White, 'The march of civilization', *Emu*, 9: 3, 1910, p. 165.

22 See, for example, Edwin Ashby, 'Notes on extinct or rare Australian birds, with suggestions as to some of the causes of their disappearance: Part 2', *Emu*, 23: 4, 1924, pp. 294–98.

23 Spencer Roberts, 'The preservation of our birds', *Emu*, 29: 3, 1930, pp. 195–200.

24 Norman Favaloro, 'Report of the Hon. State Secretary for Victoria', *Emu*, 26: 3, 1927, p. 178.

Chapter 6: Crested Shrike-tit

1 Alec Chisholm, *Mateship with Birds*, Whitcombe & Tombs, Melbourne, 1922, p. 114.

2 Ibid., p. 172.

3 C.J. Dennis, 'Introduction' to Ibid., p. 4.

4 McGregor, *Idling*, pp. 59–61.

5 Ben Eggleton, 'Obituary: L.G. Chandler', *Bird Observer*, 584, 1980, p. 74.

6 Les Chandler, *Bush Charms*, Whitcombe & Tombs, Melbourne, 1922, pp. 5, 57, 72–73. See also Mary Chandler, compiler, *'Dear Homefolks': Letters written by LG Chandler during the First World War*, Mary Chandler, Mildura, 1988.

7 Joan Beaumont, *Broken Nation: Australians in the Great War*, Allen & Unwin, Sydney, 2013.

8 Chandler, *Bush Charms*, p. 33.

9 Charles Barrett, 'Introduction' to Ibid, pp. 3–4.

10 R.T Littlejohns and S.A. Lawrence, *Birds of Our Bush, or Photography for Nature-Lovers*, Whitcombe & Tombs, Melbourne, 1920, p. 20; see also p. 56.

11 Ibid., pp. 17–18; see also p. 58.

12 Ibid., pp. 20–25.

13 Chandler, *Bush Charms*, pp. 21–22.

14 Littlejohns and Lawrence, *Birds*, pp. 38, 55, 102.

15 Ibid., pp. 54–57, 117–18, 144.

16 Chandler, *Bush Charms*, pp. 27, 30.

17 Ibid., p. 105.

18 Chisholm, *Mateship*, p. 115.

Chapter 7: Laughing Kookaburra

1 Ian Fraser and Jeannie Gray, *Australian Bird Names: Origins and meanings*, CSIRO Publishing, Melbourne, 2nd edn, 2019, p. 137.

2   Leach, *Bird Book*, p. 7.
3   A.G. Campbell, 'A dichotomous key to the birds of Australia', *Emu*, 5: Supp. 1, 1905, p. 4.
4   J.A. Leach, *A Descriptive List of the Birds Native to Victoria, Australia*, Government Printer, Melbourne, 1908, p. 4.
5   Cayley, *What Bird?* p. xvii.
6   Dunlap, *In the Field*, pp. 40-63.
7   Tom Iredale, 'The Passing of Campbell and Leach', *Australian Zoologist*, 6: 2, 1930, p. 177; Charles Bryant, 'Introduction' to J.A. Leach, *An Australian Bird Book: A complete guide to the birds of Australia*, Whitcombe & Tombs, Melbourne, 9th ed., 1958, p. vi.
8   Penny Olsen, *Cayley & Son: The life and art of Neville Henry Cayley and Neville William Cayley*, NLA Publishing, Canberra, 2013.
9   '[Review:] *An Australian Bird Book*', *Emu*, 10:5, 1911, p. 349.
10   'Some notes by Dr. Brooke Nicholls', appended to R.H. Croll, 'Obituary: Dr. John Arthur Leach', *Emu*, 29: 3, 1930, p. 232.
11   Arnold McGill, 'Keith Hindwood's contribution to New South Wales ornithology', *Australian Birds*, 15: 1, 1980, p. 3.
12   Publisher's note on Neville W. Cayley, *What Bird Is That?*, revised signature edition, Australia's Heritage Publishing, Sydney, 2011. https://catalogue.nla.gov.au/Record/5037239?lookfor=title:(what%20bird%20is%20that)%20%23[format:Book]&offset=4&max=52944
13   Keith Hindwood, ''The Late Neville W. Cayley: An appreciation', *Emu*, 50: 1, 1950, p. 55.

Chapter 8: Paradise Parrot

1   Russell McGregor, 'Alec Chisholm and the extinction of the Paradise Parrot', *Historical Records of Australian Science*, 32, 2021, pp. 156–67.
2   Alec Chisholm, 'Seeking rare parrots', *Emu*, 24: 1, 1924, p. 30.
3   'Cyril Henry Harvey Jerrard's unpublished article "Paradise Parrot", 1924, https://webarchive.nla.gov.au/awa/20141216114710/http://www.nla.gov.au/pub/paradiseparrot/B3.html
4   Cyril Jerrard to Alec Chisholm, 31 August 1929, Chisholm Papers, box 4976.
5   International Union for the Conservation of Nature: Red List: Paradise Parrot, *Psephotellus pulcherrimus*. https://www.iucnredlist.org/species/22685156/93061054#assessment-information
6   Ewin Ashby, 'Private collections and permits', *Emu*, 22: 3, 1923, pp. 210–16.
7   Alec Chisholm, 'Private collecting—A criticism', *Emu*, 22: 4, 1923, pp. 311–15.

8   A.J. Campbell, 'The Elegant and Rock Parrots', *Emu*, 27: 3, 1928, p. 137.

9   Edwin Ashby, 'Two Neophema Parrots', *Emu*, 27: 1, pp. 1–2.

10 Penny Olsen, *Glimpses of Paradise: The quest for the Beautiful Parrakeet*, NLA Publishing, Canberra, 2007, pp. 167–73.

11 Barrow, *Passion*, pp. 175–80.

12 Roger Tory Peterson, *A Field Guide to Western Birds*, Houghton Mifflin, Boston, 1941, p. xviii.

13 'Jerrard's unpublished article "Paradise Parrot"'.

14 Chisholm, *Mateship*, p. 186.

15 Alec Chisholm, 'Discovery of the Paradise Parrot', *Queenslander*, 12 August 1922, p. 44.

16 Chisholm, *Mateship*, p. 188.

17 Dunlap, *In the Field*, p. 107.

18 Alec Chisholm to Jim Bravery, 3 March 1965, Chisholm Papers, box 5000.

19 Graham Pizzey interviewed by Gregg Borschmann for 'The People's Forest' Oral History Project, March 1995, NLA TRC 2845/67, cassette 1, side 2; transcript in Pizzey Papers.

20 Pat Comben, personal communication.

21 Greg Roberts, 'The Paradise Parrot and Eric Zillmann', http://sunshinecoastbirds.blogspot.com/2011/07/paradise-parrot-and-eric-zillman.html

Chapter 9: Noisy Scrub-bird

1   Quoted in Hubert Whittell, 'A Review of the Work of John Gilbert in Western Australia, part 4', *Emu*, 51: 1, 1951, p. 24.

2   G.T. Smith, 'Habitat use and management for the Noisy Scrub-bird *Atrichornis clamosus*', *Bird Conservation International*, 6, 1996, p. 34.

3   A.J. Campbell, 'A Rara Avis', *Australasian*, 27 November 1920, p. 57.

4   Gregory Mathews, *The Birds of Australia*, vol.8, part 1, Witherby & Co., London, 1920, p.30.

5   Edwin Ashby, 'Notes on the supposed "extinct" birds of the south-west corner of Western Australia', *Emu*, 20: 3, 1921, p. 124.

6   James Pollard, 'The Normalup camp-out', *Emu*, 27: 3, 1928, p. 165.

7   Quoted in Alec Chisholm, 'The story of the Scrub-birds: Part 2', *Emu*, 51: 3, 1951, p. 288.

8   Hubert Whittell, 'The Noisy Scrub-bird (*Atrichornis clamosus*)', *Emu*, 42: 4, 1943, pp. 217–34.

9   Serventy and Whittell, *Handbook*, pp. 239, 241.

10 A.J. Campbell, 'Missing birds', *Emu*, 14: 3, 1915, p. 167; Campbell, 'Reminiscences of a field collector', *Emu*, 15: 4, 1916, p. 251.

11 Edwin Ashby, 'Some unsolved problems of Australian avifauna', *Emu*, 26: 3, 1927, pp. 158–61.

12 Keith Hindwood, 'The Bustard', *Emu*, 38: 4, 1939, p. 417.

13 C.H.H. Jerrard, 'The rare Black-breasted Quail', *Emu*, 25: 4, 1926, pp. 287–88.

14 Penny Olsen, *Night Parrot: Australia's most elusive bird*, NLA Publishing, Canberra, 2018.

15 Dom Serventy to Alec Chisholm, 4 May 1948, Chisholm Papers, box 4993.

16 Alec Chisholm, 'Meditations on "Mystery Birds", *Wild Life*, 12: 7, 1950, p. 298.

17 Chisholm, 'Scrub Birds', p. 289.

18 Correspondence, May–June 1954, Vincent Serventy Papers, box 78.

19 Norman Robinson, 'Obituary: Hargreaves (Harley) Ogilvey Webster', *Bird Observer*, 709, 1991, p. 53.

20 Harley Webster, 'Re-discovery of the Noisy Scrub-bird, *Atrichornis clamosus*', *The Western Australian Naturalist*, 8: 3, 1962, p. 57.

21 Unless otherwise referenced, this and the following three paragraphs are based on Allan Burbidge and Eleanor Russell, 'A History of Ornithology in Western Australia', in William Davis, Walter Boles and Harry Recher (eds), *Contributions to the History of Australasian Ornithology*, vol. 4, Nuttall Ornithological Club, Cambridge, Mass., 2017, pp. 506–514, plus personal communication with Allan Burbidge and Alan Danks in October 2022.

22 Robin, *Flight*, pp. 259–60.

23 Dom Serventy to Alec Chisholm, 24 June 1966, Chisholm Papers, box 4993.

24 International Union for the Conservation of Nature: Red List: Noisy Scrub-bird, *Atrichornis clamosus* https://www.iucnredlist.org/species/22703612/213149507

Chapter 10: Sarus Crane

1 Fred T.H. Smith, 'The finding and consequent identification of the Sarus Crane in Australia', *Bird Observer*, 471, 1971, p. 7.

2 Billie Gill to Francis Ratcliffe, 11 January 1967, Papers of Francis Ratcliffe, NLA MS 2493, box 7.

3 Ibid.

4 Ibid.

5 Frances Ratcliffe to Billie Gill, 23 January 1967, Ratcliffe Papers, box 7.

6 Billie Gill to Francis Ratcliffe, 29 January 1967, Ratcliffe Papers, box 7.

7 H.B. Gill, 'First record of the Sarus Crane in Australia', *Emu*, 69: 1, 1969, pp. 49–52.

8   J.A. Bravery, 'Sarus Crane in north-eastern Queensland', *Emu*, 69: 1, 1969, pp. 52–53.

9   H.B. Gill, 'Further records of Sarus Crane in northern Queensland', *Emu*, 71: 3, 1971, pp. 140–41.

10 Fred T.H. Smith, 'Birds of the month: cranes: Brolga and Sarus Crane', *Bird Observer*, 695, 1990, p. 16.

11 See, for example, Alec Chisholm, 'Some VIPs fly in from India', *Sydney Morning Herald*, 16 September 1967, p. 17.

12 Richard Schodde, 'New subspecies of Australian birds', *Canberra Bird Notes*, 13: 4, 1988, p. 119.

13 Sue Taylor, *How Many Birds is That? From the Forty-spotted Pardalote on Bruny Island to the White-tailed Tropicbird on Cape York*, Hyland House, Melbourne, 2001, pp. 121–22.

14 Danny Rogers, 'Twitchers' pilgrimage', *Wingspan*, 12: 2, 2002, p. 30.

15 John Grant, 'Secrets of the Sarus Crane', *Wingspan*, 14: 4, 2004, pp. 18–19; 'Letters to the editor', *Wingspan*, 15: 3, 2005, pp. 42–43.

16 Billie Gill, John Grant, Graham Harrington, Elinor Scambler and Virginia Simmonds, 'The mysterious affair of the Sarus Crane', *Australian Birdlife*, 2014, 3: 1, p. 34.

17 Timothy Nevard et al, 'Subspecies in the Sarus Crane *Antigone antigone* revisited: with particular reference to the Australian population', *PLoS ONE*, 15: 4, 2020. https://doi.org/10.1371/journal.pone.0230150

18 G.R. Beruldsen, 'Is the Sarus Crane under threat in Australia', *Sunbird*, 27: 3, 1997, p. 74.

19 Peter Sutton, *Wik-Ngathan Dictionary*, Caitlin Press, Adelaide, 1995, pp. 26, 110, 139. See also Philip Clarke, *Aboriginal Peoples and Birds in Australia: Historical and cultural relationships*, CSIRO Publishing, Melbourne, 2023, pp. 90, 93.

20 Donald Thomson, *Birds of Cape York Peninsula: Ecological notes, field observations, and catalogue of specimens collected on three expeditions to north Queensland*, Government Printer, Melbourne, 1935.

21 Beruldsen, 'Sarus Crane', p. 75.

22 'Fred Smith again', *Bird Observer*, 423, 1967, p. 3.

23 'The Wingspan Interview: Fred Smith', *Wingspan*, 19: 4, 2009, p. 58.

24 Sean Dooley, 'From the Editor', *Wingspan*, 19: 4, 2009, p. 5.

25 Slater, *Birdwatcher's Notebook*, pp. 84–88.

26 Norman Favaloro and Allan McEvey, 'A new species of Australian grass-wren', *Memoirs of the National Museum of Victoria*, 28, 1968, pp. 1–9.

27 *Bird Observer*, May–August 1962.

28 H.E.A. J[arman], 'Book review', *Bird Observer*, 527, 1975, pp. 76–77.

29 *Bird Observer*, 544, March 1977, p. 17; 551, October 1977, p. 81; 553/4,

December 1977/January 1978, p. 96.

30 *Wingspan*, 9, March 1993, p. 22; 10, June 1993, p. 28; 12, December 1993, p. 26; 13, March 1994, p. 14.

31 Schodde, 'New subspecies', p. 119.

32 Stephen Garnett and Gabriel Crowley, 'The History of Threatened Birds in Australia and its Offshore Islands' in William Davis, Harry Recher, Walter Boles and Jerome Jackson, *Contributions to the History of Australasian Ornithology*, Nuttall Ornithological Club, Cambridge, Mass, 2008, pp. 415–416; '*Grus antigone* Sarus Crane', *Handbook of Australian, New Zealand and Antarctic Birds*, Oxford University Press, Melbourne, vol. 2, 1993, p. 482.

33 Australian Crane Network: https://ozcranes.net/species/sarus.html

34 Elinor Scambler, Timothy Nevard, Graham Harrington, E. Ceinwen Edwards, Virginia Simmonds and Donald Franklin, 'Numbers, distribution and behaviour of Australian Sarus Cranes *Antigone antigone gillae* and Brolgas *A rubicunda* at wintering roosts on the Atherton Tablelands, far north Queensland, Australia', *Australian Field Ornithology*, 37, 2020, pp. 87–99.

35 Crane counts and teams: https://ozcranes.net/research/escamb1_1. html

Chapter 11: Rock Warbler

1 Keith Hindwood, 'The Rock-Warbler: A monograph', *Emu*, 26: 1, 1926, pp. 14–24.

2 A.R. McG[ill], 'Obituary: Keith Alfred Hindwood', *Emu*, 71: 4, 1971, p. 183.

3 Norman Chaffer, 'Photographing the Rock-Warbler', *Emu*, 26: 1, 1926, p. 24.

4 Keith Hindwood, 'The Rock Warbler', *Australian Museum Magazine*, 3: 11, 1929, p. 382.

5 Alec Chisholm, *Bird Wonders of Australia*, Angus & Robertson, Sydney, 4th ed., 1956, pp. 117–18.

6 'Australian Museum Keith Hindwood Memorial Fund', *Birds*, 6: 2, 1971, p. 23.

7 Alec Chisholm, 'Keith Hindwood—a man who loved Sydney', *Australian Author*, 3: 4, 1971, p. 19.

8 Ibid., p. 18.

9 'Hindwood Memorial Fund', p. 24.

10 Allen Keast, 'The Sydney ornithological fraternity, 1930s–1950: anecdotes of an admirer', *Australian Zoologist*, 30: 1, 1995, p. 29.

11 Roy Cooper, 'Obituary: Mr. K.A. Hindwood', *Bird Observer*, 474, 1971, p. 5

12 Hindwood, 'Rock Warbler', 1929, p. 383.
13 Ibid., p. 383.
14 Penny Olsen, *An Eye for Nature: The life and art of William T. Cooper*, NLA Publishing, Canberra, 2014, pp. 29-34.
15 Slater, *Masterpieces*, p. 77.
16 Keith Hindwood, 'Strange nesting site of Magpie-Larks', *Emu*, 37: 3, 1938, p. 242.
17 McGill, 'Obituary', p. 183.
18 Cooper, 'Obituary', p. 5; Keast, 'Sydney ornithological fraternity', p. 27.
19 Robin, *Flight*, pp. 209-29.
20 Quoted in Olsen, *William Cooper*, p. 34.
21 Robin, *Flight*, pp. 223-25.
22 McGill, 'Obituary', p. 184.
23 Hindwood, 'Rock Warbler', 1929, p. 384.

Chapter 12: Collared Sparrowhawk

1   H.T. Condon, *Field Guide to the Hawks of Australia*, Bird Observers Club, Melbourne, 1949, pp. 6-7.
2   Condon, *Field Guide to the Hawks of Australia*, Bird Observers Club, Melbourne, 4th ed., 1966, pp. 8-9.
3   Menkhorst et al, *Australian Bird Guide*, pp. 232-33.
4   Dunlap, *In the Field*, pp. 42, 103.
5   Russell McGregor, 'Before Slater: A history of field guides to Australian birds to 1970', *Australian Field Ornithology*, 39, 2022, pp. 125-38.
6   Peter Slater to ACF, 12 October 1965, Ratcliffe Papers, box 5.
7   Email, Robin Hill to Russell McGregor, 23 October 2023.
8   Graham Pizzey to Francis Ratcliffe, 14 June 1965, Ratcliffe Papers, box 6.
9   Graham Pizzey, 'The Australian Goshawk', *Wild Life*, 10: 10, October 1948, p. 442.
10 Francis Ratcliffe to A. Dunbavin Butcher, 6 October 1966, Ratcliffe Papers, box 5.
11 Francis Ratcliffe to J.D. Macdonald, 13 January 1966, Ratcliffe Papers, box 5.
12 Francis Ratcliffe to Eric Lindgren, 22 February 1967, Ratcliffe Papers, box 5; Francis Ratcliffe to Graham Pizzey, 22 February 1967, Ratcliffe Papers, box 6.
13 Francis Ratcliffe to Norman Wettenhall, 12 May 1965, Ratcliffe Papers, box 7; Francis Ratcliffe to R.D. Piesse, 28 April 1969, Ratcliffe Papers, box 6.
14 Graham Pizzey to Francis Ratcliffe, 7 December 1965, Ratcliffe Papers, box 6.

15 Graham Pizzey to Francis Ratcliffe, 5 April 1966, Ratcliffe Papers, box 6.

16 Barbara and Roger Peterson's New Year's letter, 1972, Pizzey Papers.

17 Roger Peterson to Graham Pizzey, 7 May 1971, Pizzey Papers.

18 Graham Pizzey to Francis Ratcliffe, 4 April 1967, Ratcliffe Papers, box 6.

19 Pizzey interview, NLA TRC 2845/67, cassette 2, side 1, transcript p. 8. According to Robin Hill (email, 23 October 2023), the break-up was prompted by disagreements over advances on royalties.

20 Francis Ratcliffe to Graham Pizzey, 23 May 1968, Ratcliffe Papers, box 6.

21 Peter Slater to Eric Lindgren, 7 January 1968, Ratcliffe Papers, box 5. See also Eric Lindgren to H.J. Frith, 6 February 1968, Ratcliffe Papers, box 5.

22 H.E.A. J[arman], 'Book reviews', *Bird Observer*, 589, 1981, p. 8.

23 Alan Morris, 'Book review', *Australian Birds*, 15: 2, 1980, pp. 34–36

24 Lawrie Conole, 'Field guides: Which ones do Australian birders prefer?', *Galah*, 27, 1998, pp. 2–3.

25 Stephen Debus, *The Birds of Prey of Australia: A field guide to Australian raptors*, Oxford University Press, Melbourne, 1998, pp. 77–90.

26 Condon, *Hawks*, 1949, p. 2.

27 Pizzey interview, NLA TRC 2845/67, cassette 3, side 1, transcript p. 5.

28 Simon Barnes, *How to be a Bad Birdwatcher*, Short Books, London, 2004, p. 59.

Chapter 13: Jacky Winter

1   Fraser and Gray, *Bird Names*, p. 262.

2   Leach, *Bird Book*, p. 74.

3   RAOU, *Official Checklist of the Birds of Australia*, Government Printer, Melbourne, 2nd ed., 1926, pp. iv–v.

4   'Proceedings of the Annual Congress of the RAOU, Perth, 1948', *Emu*, 48: 3, 1949, p. 199.

5   Charles Bryant, 'The little bird with the long name', *Wild Life*, 10: 9, 1948, p. 403.

6   Arnold McGill, 'Review: *Checklist of the Birds of Australia, Part 1 - Non-passerines* by H.T. Condon', *Australian Bird Bander*, 13: 3, 1975, p. 66.

7   Alec Chisholm, *Nature Fantasy in Australia*, Dent, London, 1932, p. 163.

8   Dom Serventy to Richard Schodde, 24 March 1977, Pizzey Papers.

9   Alec Chisholm to Cecil Cameron, 18 April 1972, Chisholm Papers, box 4993.

10 'Recommended English Names', pp. 247, 248.

11 Serventy to Schodde, 24 March 1977, Pizzey Papers.

12 Leach, *Descriptive List*, pp. 3–4.

13 W.B. Alexander, 'Popular names for Australian birds', *Emu*, 33: 2, 1933, pp. 110–11.

14 J.A. Leach, 'The naming of Australian birds', *Emu*, 24: 3, 1925, p. 182.

15 'Recommended English Names', pp. 276, 300, 303.

16 Ibid., p. 294.

17 Marion Cassels, 'Changing bird names', *Bird Observer*, 577, 1979, p. 89.

18 Roger Thomas, 'A plea for uniformity', *Bird Observer*, 570, 1979, p. 27.

19 Nancy Barraclough, 'BOC members around the Kimberleys', *Bird Observer*, 579, 1980, p. 13.

20 H.E.A.J[arman], 'Recommended English names for Australian birds', *Bird Observer*, 561, 1978, p. 42; Howard Jarman, 'English Names for Australian Birds', *Bird Observer*, 572, 1979, p. 45.

21 Alastair Morrison, 'In support of "new" names', *Bird Observer*, 574, 1979, p. 61.

22 Allan McEvey to Graham Pizzey, 24 November 1980, Pizzey Papers.

23 Graham Pizzey, *A Field Guide to the Birds of Australia*, Collins, Sydney, 1980, p. 14.

24 Roger Jaensch, 'Book review', *South Australian Ornithologist*, 28, 1981, p. 192.

25 J. Penhallurick, 'Mixed blessings in new field guide', *Canberra Times*, 30 November 1980, p. 8.

26 Graham Pizzey to Richard Schodde, 8 September 1982, Pizzey Papers.

27 Graham Pizzey to Allen Keast, 19 September 1983, Pizzey Papers.

28 Peter Higgins, 'REN or fairy-wren? The debate continues' *Wingspan*, 13, 1994, pp. 23–26.

29 Peter Higgins 'And the winner is ... Recommended English names', *Wingspan*, 5: 1, 1995, pp. 20–23.

30 John Squire 'Our birds and our vernacular', *Wingspan*, 5: 1, 1995, pp. 25, 36.

31 Les Christidis, 'More about name calling', *Wingspan*, 5: 2, 1995, pp. 10–11.

32 See, for example, Stephen Garnett. Golo Maurer and Georgia Garrard, 'Why Australian common bird names should respond to societal change', *Emu*, 122: 2, 2022, pp. 150–152.

Chapter 14: Straw-necked Ibis

1 Henry Nix, 'The President Writes', *Wingspan*, 11: 3, 2001, p. 4.

2 Alec Chisholm, *The Joy of the Earth*, Collins, Sydney, 1969, p. 162.

3 Dorothy Kass, 'Gould League of Bird Lovers', https://dehanz.net.au/entries/gould-league-of-bird-lovers/

4 Walter Finigan, 'The Gould League of Bird Lovers of New South

Wales' in Neville Cayley (ed.), *Feathered Friends: A Gould League annual*, Angus & Robertson, Sydney, 1935, p. 9.

5   Patrick Bourke, 'My blarney birds', *Wild Life*, 10: 12, 1948, p. 552.

6   *Gould League Songs & Poems*, New South Wales Gould League of Bird Lovers, Sydney, 1934, p. 3.

7   R.P. McLellan, 'Foreword', *Bird Lover*, 15, 1962, p. 2.

8   Finigan, 'Gould League', p. 11.

9   Dorothy Kass, *Educational Reform and Environmental Concern: A history of school nature study in Australia*, Routledge, London, 2018, p. 172.

10   See, for example, F.G. Elford, 'Hints on bird-watching', *Bird Lover*, 2, 1949, pp. 7–8; *How to Study Birds: A Gould League publication*, Paterson Brokensha, Perth, 1953.

11   'Allambee South observations', *Bird Lover*, 16, 1963, pp. 7–9.

12   Virginia Jackson and Glenda Brigham, 'Bird excursion to Toolern Vale', *Bird Lover*, 12, 1959, pp. 6–8.

13   H.N. Beck, 'Origin of the Gould League of Bird Lovers', *Bird Lover*, 6, 1953–54, p. 19.

14   David Street, 'The Straw-necked Ibis', *Bird Lover*, 6, 1953–54, p. 26.

15   Drawings by Kerin Swinburn, *Bird Lover*, 12, 1959, p. 21.

16   Roy Wheeler, 'Straw-necked Ibis', *Bird Lover*, 17, 1964, p. 31.

17   'President's Address: The most useful bird in Australia', *Emu*, 8: 3, 1909, p. 169.

18   Leach, *Bird Book*, p. 54.

19   Perrine Moncrieff, 'Australian White Ibis in New Zealand', *Emu*, 25: 4, 1926, pp. 296–97.

20   'Haunts of the ibis', *Wild Life*, 3: 2, 1941, p. 72.

21   Belcher, *Birds of Geelong*, p. 106.

22   Colin Grant, 'The Straw-necked Ibis', *Bird Lover*, 15, 1962, p. 9.

23   *Bird Lover*, 22, 1969.

24   C.R. Forster, 'Straw-necked Ibis', *Gould League Notes 1938*, p. 36; copy courtesy of Emily Gallagher.

25   ABC News: https://www.abc.net.au/news/2017-12-11/bird-of-the-year-magpie-defeats-teambinchicken/9245242

26   Darryl Jones, *Curlews on Vulture Street: Cities, birds, people and me*, NewSouth, Sydney, 2023, pp. 273–95.

27   https://theconversation.com/friday-essay-the-rise-of-the-bin-chicken-a-totem-for-modern-australia-100673

28   https://www.feathersandphotos.com.au/phpbb/ibis-washing-cane-toad-t45204.html

29   https://www.973fm.com.au/lifestyle/this-just-in-ibis-have-a-purpose-in-society/

30 'How "bin chickens" learnt to wash poisonous cane toads', https://www.bbc.com/news/world-australia-63699884

Chapter 15: Blue-faced Parrot-Finch

1   G.A. Heuman, 'Parrot-Finches (Erythrura)', *Emu*, 25: 2, 1925, p. 49.
2   Neville W. Cayley, *Australian Finches in Bush and Aviary*, Angus & Robertson, Sydney, 1932, p. 108.
3   Klaus Immelmann, *Australian Finches in Bush and Aviary*, Angus & Robertson, Sydney, 1965, pp. vii, 76.
4   Moss, *Bird in Bush*, p. 265; Mark Cocker, *Birders: Tales of a Tribe*, Vintage, London, 2002, pp. 52–53.
5   *Bill Oddie's Little Black Bird Book*, Robson, London, 2006 [1980], p. 22.
6   Dunlap, *In the Field*, pp. 57–59, 72–83; Weidensaul, *Of a Feather*, pp. 184, 205; Moss, *Bird in Bush*, p. 265.
7   Gillian Lord, 'History of ornithology in Tasmania', *Tasmanian Historical Research Association: Papers and Proceedings*, 33: 3, 1986, p. 101.
8   Rosemary Balmford, *Learning about Australian Birds*, Collins, Sydney, 1980, p. 111.
9   Rosemary Balmford, 'People and birds in Australia 1980–1990', *Bird Observer*, 717, 1992, p. 11.
10 Slater, *Birdwatcher's Notebook*, p. 51.
11 Graham Pizzey, 'On being a birdwatcher', *Bird Observer*, 726, 1992, p. 6.
12 Andrew Stafford, '[review:] *The Field Guide to the Birds of Australia*', *Australian Birding*, 3: 4, 1997, p. 22.
13 Andrew Stafford, 'The original great grasswren expedition, 7–31 August 1991', *Australian Birding*, 2: 1, 1995.
14 Cocker, *Birders*, p. 54.
15 Balmford, 'People and birds', p. 11.
16 Sean Dooley, 'A twitch in time: 25 years of the Twitchathon', *Wingspan*, 18: 4, 2008, pp. 38–39.
17 Roy Wheeler, 'The Bird Count Challenge', *Bird Observer*, 438, 1968, pp. 6–7.
18 Roy Wheeler, 'Scoring a century', *Wild Life*, 8: 12, 1946, pp. 431–35.
19 Roy Wheeler, 'Birds seen at a test match at Melbourne', *Bird Observer*, 411, 1966, p. 7.
20 Michael Sharland, *Birds of the Sun*, Angus & Robertson, Sydney, 1967, p. 177.
21 W.B. Alexander to Alec Chisholm, 26 Feb 1923, Chisholm Papers, box 4988.
22 Alec Chisholm, 'Seeking "new" birds', *Argus*, 1 April 1933, p. 4.
23 'Twitchathon: Twitching for conservation', *Wingspan*, 8: 1, 1998, p. 27.

24 'Whither Australia?' *Australian Birding*, 1: 1, [1994]; 'The great debate', *Australian Birding*, 1: 2, 1994.
25 Mike Carter, 'Obituary: John Leonard McKean', *Wingspan*, 6: 2, 1996, p. 29; Hilary Thompson, 'The list and times of John McKean', *Australian Birding*, 2: 4, 1995–96, p. 7.
26 Ken Simpson and Michael Weston, 'BOCA celebrates 100 years', *Wingspan*, 15: 4, 2005, p. 5.
27 Cocker, *Birders*, p. 55.
28 Taylor, *How Many*, p. v.

Chapter 16: Eungella Honeyeater
1   J.S. Robertson, 'Mackay Report', *Emu*, 61: 4, 1962, pp. 272–73.
2   Alec Chisholm, 'Bird-notes from Eungella', *Emu*, 65: 3, 1965, p. 164.
3   Alec Chisholm, 'The Diaries of S.W. Jackson', *Emu*, 58: 2, 1958, p. 104.
4   Email, Walter Boles to Russell McGregor, 7 December 2022.
5   Wayne Longmore and Walter Boles, 'Description and systematics of the Eungella Honeyeater, *Meliphaga hindwoodi*, a new species of honeyeater from central eastern Queensland, Australia', *Emu*, 83: 2, 1983, pp. 59–60.
6   Jock Marshall, field notes, July–October 1930, Papers of Jock Marshall, NLA MS 7132, box 11, item 3/2.
7   Longmore and Boles, 'Eungella Honeyeater', p. 59.
8   R.L. Kitching, 'Eungella—The Land of Clouds Revisited', *Proceedings of the Royal Society of Queensland*, 125, 2020, p. 1.
9   Fraser and Gray, *Bird Names*, pp. 143–44.
10 Edward Sorenson, 'Aboriginal names of birds', *Emu*, 20: 1, 1920, pp. 32–33.
11 Ian Abbott, 'Aboriginal names of bird species in south-west Western Australia, with suggestions for their adoption into common usage', *Conservation Science Western Australia*, 7: 2, 2009, p. 215.
12 Nick Thieberger and William McGregor (eds), *Macquarie Aboriginal Words*, Macquarie Library, Sydney, 1994, pp. 50, 90–91.
13 Les Christidis and Walter Boles, *The Taxonomy and Species of Birds of Australia and its Territories*, RAOU, Melbourne, 1994, p. 22.
14 Árpád Nyári and Leo Joseph, 'Systematic dismantlement of *Lichenostomus* improves the basis for understanding relationships within the honeyeaters (Meliphagidae) and the historical development of Australo-Papuan bird communities', *Emu*, 111: 3, 2011, pp. 207–08.
15 Higgins, 'REN or fairy-wren?', p. 23.
16 Barrett, *Koonwarra*, p. 161.
17 Pizzey, *Field Guide*, 1980, p. 13.
18 Leach, *Bird Book*, p. 119.

Chapter 17: Common Myna

1 Hugo Phillipps, 'Cane toads with wings', *Wingspan*, 13, 1994, pp. 11–12.

2 'Bird conservation: Facing the future', *Wingspan*, 9: 4, 1999, p. 14,

3 E.L. Jones, letter to editor, *Wingspan*, 14, 1994, pp. 26–27.

4 Heather Gibbs, 'Indian Mynas', *Wingspan*, 15, 1994, p. 28.

5 See, for example, Nicholas Smith, 'Thank you mother for the rabbits: Bilbies, bunnies and redemptive ecology', *Australian Zoologist*, 33: 3, 2006, pp. 369–78; David Trigger, Jane Mulcock, Andrea Gaynor, and Yann Toussaint, 'Ecological restoration, cultural preferences and the negotiation of "nativeness" in Australia', *Geoforum*, 39, 2008, pp. 1273–83.

6 Ken Simpson, '"What are the greatest threats to birdlife in Victoria at present, and what steps can be taken to ease them?"', *Bird Observer*, 518, 1975, p. 2.

7 Robert Hall, *The Useful Birds of Southern Australia: with notes on other birds*, Lothian, Melbourne, 1907, p. 299.

8 Froggatt, *Useful Birds*, pp. 18, 36.

9 Lucas and Le Souëf, *Birds*, pp. 436–40.

10 See, for example, A.H. Wilson, 'An impression of the birds of Australia', *Emu*, 21: 3, 1922, p. 202.

11 Dom Serventy, 'The menace of acclimatization', *Emu*, 36: 3, 1937, p. 189.

12 'Birds and the evacuation: An ornithologist's view', *Wild Life*, 4: 9, 1942, pp. 331–32.

13 'Proceedings of the 35th Annual Congress of the RAOU, Adelaide, 1936', *Emu*, 36: 3, 1937, pp. 167–68

14 Charles Bryant, 'Friendly alien', *Wild Life*, 6: 7, 1944, pp. 199–200.

15 Crosbie Morrison, 'Backyard diary', *Argus*, 8 July 1955, p. 4

16 Alec Chisholm, 'Sydney birds ring changes', *Sydney Morning Herald*, 7 November 1964.

17 Alec Chisholm, *Bird Wonders of Australia*, Angus & Robertson, Sydney, 1934, pp. 225–27.

18 Alec Chisholm, 'Hustle and bustle birds', *Sydney Morning Herald*, 13 February 1971.

19 McGregor, *Idling*, pp. 93–94, 108–09.

20 Ted Schurmann, *Bird-watching in Australia*, Rigby, Adelaide, 1977, pp. 70–73.

21 Peter Trusler, Tess Kloot, and Ellen McCulloch, *Birds of Australian Gardens*, Rigby, Adelaide, 1980, pp. 94, 162, 170-171.

22 Clifford Frith, *Garden Birds: Attracting birds to Australian and New Zealand gardens*, Doubleday, Sydney, 1985, p. 144.

23 Rosemary Balmford, 'Introduced birds in Australia', *Bird Observer*, 671, 1988, p. 3.

24 Glen Johnson, 'Letter to the editor – introduced birds', *Bird Observer*, 675, 1988, p. 59.

25 Quoted in Noel Luff, 'A change in attitude: where did the Indian Myna go wrong?' *Canberra Bird Notes*, 41: 2, 2016, p. 121.

26 Canberra Indian Myna Action Group, https://indianmynaaction.org.au/

27 Luff, 'Change in attitude', p. 121.

Chapter 18: Rainbow Lorikeet

1   E.M. Cornwall, 'Wild parrot pets', *Emu*, 10: 2, 1910, pp. 135–36.

2   *Wild Life*, 9: 4, 1947, p. 126.

3   Frank Robinson, 'In favour of a bird table', *Wild Life*, 4: 10, 1942, pp. 370–72.

4   Schurmann, *Bird-watching*, pp. 112-113; Ellen McCulloch, 'Feeding of birds', *Bird Observer*, 541, 1976, pp. 89–90.

5   https://www.sydneywildlife.org.au/post/what-can-i-feed-the-birds-in-my-garden

6   https://www.environment.nsw.gov.au/questions/neighbour-feeding-birds

7   Darryl Jones, *The Birds at my Table: Why we feed wild birds and why it matters*, NewSouth, Sydney, 2018; Jones, *Feeding the Birds at your Table: A guide for Australia*, NewSouth, Sydney, 2019.

8   Jones, *Curlews*, pp. 244–45.

9   S.R. White, 'A bird bath and feeding table', *How to Study Birds*, [1953], p. 13.

10 Elford, 'Hints', p. 8.

11 J.D. Jennings, 'Birds and children', *Emu*, 27: 1, 1927, p. 52.

12 John Rowland Skemp, 'My friend Midgie', *Wild Life*, 11: 8, 1949, pp. 347–48, 380, 383.

13 John Rowland Skemp, *My Birds*, C.L. Richmond & Sons, Devonport, [1970], pp. 45-57.

14 Littlejohns and Lawrence, *Birds*, p. 103.

15 Lucy Lee, 'You can't beat a spade', *Wild Life*, 10: 11, 1948, pp. 490–92.

16 Patrick Bourke, *A Handbook of Elementary Bird Study*, Paterson Brokensha, Perth, 1955, p. 116.

17 Balmford, *Learning*, pp. 98–99; Balmford, *The Beginner's Guide to Australian Birds*, Penguin, Melbourne, rev. edn, 1990, pp. 99-100.

18 Jane Fletcher, *Tasmania's Own Birds*, Mercury Press, Hobart, [195?], pp. 21–30.

19 See, for example, Barbara Salter, 'Food for the Winter', *Bird Observer*, 426, 1967, p. 4; Salter, *Australian Native Gardens and Birds*, Jacaranda, Brisbane, 1969, pp. 18–29.

20   Florence Vasey, 'Artificial feeding of honeyeaters'; *Bird Observer*, 456, 1969, pp. 4–5; Barbara Salter, 'Artificial feeding of honeyeaters', *Bird Observer*, 457, 1969, p. 3.

21   Ellen McCulloch, 'Food for thought', *Bird Observer*, 636/7, 1985, p. 12.

22   Kloot, McCulloch, and Trusler, *Birds*, pp. 16–17.

23   Graham Pizzey, *A Garden of Birds: Australian birds in Australian gardens*, Viking O'Neil, Melbourne, 1988, pp. 10–17.

24   Simpson and Wilson, *Birdwatching*, pp. 9–11, 54.

25   Jones, *Feeding the Birds*, p. 18.

26   '50 things you can do to help birds, birding and Birds Australia' *Wingspan*, 10: 1, 2000, p. 24.

27   Michelle Plant 'Good practice when feeding wild birds', *Wingspan*, 18: 1, 2008, pp. 20–23.

28   Darryl Jones, 'Feed the birds? The need for a guide to guilt-free bird feeding', *Wingspan*, 18: 1, 2008, pp. 16–19.

29   David Andrew, *Pocket Garden Birdwatch*, Dorling Kindersley, Melbourne, 2015, pp. 42–47.

30   Jones, *Birds at my Table*, pp. 257–62, 277–81.

31   Peter Slater, Raoul Slater, and Sally Elmer, *Visions of Wildness*, Reed New Holland, London, 2016, pp. 15, 280–81.

Chapter 19: Golden-shouldered Parrot

1   Russell McGregor, *Environment, Race, and Nationhood in Australia: Revisiting the empty north*, Palgrave Macmillan, New York, 2016; Tim Rowse, 'Indigenous heterogeneity', *Australian Historical Studies*, 45: 3, 2014, pp. 297–310.

2   Gabriel Crowley and Stephen Garnett, 'Distribution and decline of the Golden-shouldered Parrot *Psephotellus chrysopterygius* 1845-1990', *North Queensland Naturalist*, 53, 2023, pp. 22–68.

3   *Antbed: An occasional newsletter about the Golden-shouldered Parrot produced by Stephen Garnett and Gabriel Crowley*, 1, c. February 1993 and 2, July 1993; Stephen Garnett, 'Are you visiting Cape York?' *Bird Observer*, 728, March 1993, p. 7.

4   Garnett, 'Are you visiting?'.

5   Email, Stephen Garnett to Russell McGregor, 17 October 2022.

6   *Antbed*, 6, October 1995, pp. 14–15.

7   *Antbed*, 1, p. 1.

8   *Antbed*, 3, April 1994, p. 3.

9   *Antbed*, 6.

10   Ibid., p. 12.

11   Golden-shouldered Parrot Recovery Team, *Draft Golden-shouldered*

Parrot Recovery Plan, Olkola Aboriginal Corporation, Cairns, 2022, p. vii.

12   Ibid., pp. vi–vii, 14.

13   'BirdLife Australia's mission to save birds': https://birdlife.org.au/our-impact/

14   Dunlap, *In the Field*, pp. 149–59; Thomas Dunlap, *Nature and the English Diaspora: Environment and history in the United States, Canada, Australia, and New Zealand*, Cambridge University Press, Cambridge, 1999, pp. 275–305.

15   Stephen Garnett and Gabriel Crowley, *Recovery Plan for the Golden-shouldered Parrot Psephotus chrysopterygius 2003-2007*, Queensland Parks and Wildlife Service, Brisbane, 2002, pp. 4, 7.

16   Steve Murphy, Susan Shephard, Gabriel Crowley, Stephen Garnett, Patrick Webster, Wendy Cooper and Rigel Jensen, *Pre-management Actions Baseline Report for Artemis Antbed Parrot Nature Refuge*, Threatened Species Recovery Hub, Brisbane, 2021; 'Artemis Nature Fund: 2021 in review': https://mailchi.mp/cfee315bc38c/artemis_2021_summary

17   Anthony Ham, 'Australia's most endangered parrot faces an unusual threat: trees', *Smithsonian Magazine*, 7 February 2023, https://www.smithsonianmag.com/science-nature/australias-golden-shouldered-parrot-faces-an-unusual-threat-180981591/

Chapter 20: Regent Bowerbird

1   John Bransbury, *Where to Find Birds in Australia*, Century Hutchinson, Melbourne, 1987, p. xiv.

2   Peter O'Reilly, *25 Years of Bird Week*, Brisbane, c. 2003.

3   David Callahan, *A History of Birdwatching in 100 Objects*, Bloomsbury, London, 2014, p. 100.

4   Campbell, 'Camp-out Phillip Island', p. 208.

5   S.A. White, 'Narrative of the expedition promoted by the Australasian Ornithologists' Union to the islands of Bass Strait', *Emu*, 8: 4, 1909, pp. 195–207.

6   G.H. Barker, 'Narrative of proceedings at the 23rd congress and camp of the RAOU', *Emu*, 24: 3, 1925, pp. 208–17; Alec Chisholm, 'The Yeppon [sic]-Byfield excursion', *Emu*, 24: 3, 1925, pp. 221–29.

7   Arnold McGill, 'Jack Ramsay: A pioneer ornithological photographer', *Australian Bird Watcher*, 9: 5, 1982, p. 170.

8   *Emulet*, 1, January 1949, pp. 8–9, 11.

9   Robin, *Flight*, p. 137.

10 'Yearly holiday outings??', *Bird Observer*, 452, 1969, p. 2.

11 'BOC on Cape York in 1970', *Bird Observer*, 455, 1969, p. 2.

12 'Another BOC outing in 1970', *Bird Observer*, 456, 1969, p. 2.

13 'Trip report: Soviet Central Asia bird tour May/June 1986', *Bird Observer*, 656, 1986, np.

14 David Andrew, 'Birding at Iron Range, Qld', *Bird Observer*, 659, 1986, p. 132.

15 Bransbury, *Where to Find*, p. vii.

16 Richard Thomas, Sarah Thomas, David Andrew and Alan McBride, *The Complete Guide to Finding the Birds of Australia*, CSIRO Publishing, Melbourne, 2nd ed., 2011, pp. viii, 70.

17 Darrell Price, 'The birds of Cape York Wilderness Lodge', *Bird Observer*, 678, 1988, pp. 85–86.

18 Susan Bailey, 'A bird-watching holiday in eastern Australia', *Bird Observer*, 697, 1990, pp. 44–45; Connie and John Trotter, 'Good birding along the south Queensland coast', *Bird Observer*, 515, 1974, pp. 10–11.

19 O'Reilly, *Bird Week*, p. 2.

20 Peter O'Reilly, *The Spirit of O'Reillys: The world at our feet*, Peter O'Reilly, [Brisbane?], 2008, p. 222.

Chapter 21: Superb Fairy-wren

1 Birds in Backyards: https://www.birdsinbackyards.net/Achievements-and-History

2 Aussie Bird Count: https://aussiebirdcount.org.au/

3 Harry Wolstenholme, 'Notes from Wahroonga, Sydney, NSW', *Emu*, 22: 2, 1922, pp. 141–48.

4 Harry Wolstenholme, 'Nesting notes from a Sydney garden', *Emu*, 28: 3, 1929, p. 188.

5 Chisholm, *Nature Fantasy*, pp. 18–39.

6 McGregor, *Idling*.

7 Gay Grogan, 'Backyard birding, Croydon, Vic.', *Bird Observer*, 641, 1985, p. 60.

8 Barbara Burns 'Birds on busy schedule', *Bird Observer*, 754, 1995, p. 8.

9 Molly Brown, 'Our resident swallows', *Bird Observer*, 752, 1995, pp. 8–9; Brown, 'Roadside walk', *Bird Observer*, 764, 1996, p. 6; Brown, 'Watching at the window', *Bird Observer*, 755, 1995, pp. 6–7.

10 Arnold McGill, 'Records, reminiscences and reflections', *Bird Observer*, 582, 1980, pp. 49–50.

11 Sue Taylor, *The 100 Best Birdwatching Sites in Australia*, John Beaufoy, Oxford, 2021, p. 15.

12 Charles Bryant, 'Birds of the bend: Beauty in a municipal garbage tip', *Wild Life*, 9: 8, 1947, pp. 295–98.

13 Pizzey interview, NLA TRC 2845/67, cassette 1, side 2, transcript p. 6.

14 Sean Dooley, *Anoraks to Zitting Cisticolas: A whole lot of stuff about birdwatching*, Allen & Unwin, Sydney, 2007, pp. 1-2.

15 Neville W. Cayley, 'A blue bird of happiness', in Cayley (ed.), *Feathered Friends*, p. 51.

16 Jeremy Mynott, *Birdscapes: Birds in our imagination and experience*, Princeton University Press, Princeton, 2009, p. 23.

17 Trusler, Kloot, and McCulloch, *Birds*, p. 11.

Index